Chemistry
Practical Series

This page is left intentionally blank

Volumetric Analysis:
Concepts and Experiments

Dr. Chirag Fultariya & Dr. Jalpa Harsora

2017

Title of the Book: Volumetric Analysis: Concepts and Experiments

ISBN: 978-1-365-79930-3

First Printing: March 2017

To
Our Mother
in
Heaven

INDEX

Preface

Volumetric Analysis requires special attention when students are of entry level. Laboratory setup of these analysis and the basic methodology involve is among the most important things for entry level students.

Here we have try to simplify the things. So that students can understand basic things about Volumetric Analysis which will be very helpful to them in future.

This book mainly focus on Volumetric Analysis but it also contains ample information of safety and glasswares used in a laboratory. Safety is a major concern when we are dealing with a chemical. It is necessary to have some common information and knowledge while dealing with chemicals.

Glasswares are key component of any experiments performed in laboratory. In-depth understanding about the use of these glassware is must. In this book I have try to summarize some common glasswares that are used in laboratory during inorganic qualitative analysis.

Volumetric analysis (also determination or titration) is generally one of the quantitative methods that deal with determination of quantity (amount) of particular elements (components) in tested (analyzed) material (sample). It is preceded by qualitative analysis, when only composition of analyzed sample is tested.

I hope that this book will surly help students to get good command over Volumetric Analysis in their laboratory practice and will help them to score in chemistry.

Dr. Chirag R. Fultariya
&
Dr. Jalpa P. Harsora

SAFETY POINTS

A First Aid Box should be kept in a readily accessible place in the laboratory, and should contain the following articles. All bottles and packages should be clearly labelled. "Bandages, lint gauze, cotton wool, adhesive plaster, and a sling. Delicate forceps, needles, thread, safety pins, and scissors, fine glass dropping tube, Glass eye-bath, Vaseline, salt, mustard (powder), Castor oil, Olive oil, Sal volatile, Zinc Oxide ointment, Boric acid (powder), Chloramine-T (fine hydrated crystals, not as pellets)".

Bottles:
- ✓ Actiflavine or Euflavine Emulsion (in quantity).
- ✓ Saturated aqueous picric acid solution (in quantity).
- ✓ Lime water (in quantity).
- ✓ 2% Iodine solution, 1% Boric acid, 1% Acetic acid.
- ✓ 8% Sodium bicarbonate solution (i.e., saturated in the cold).
- ✓ 1% Sodium bicarbonate solution.

BURNS:

[A] Caused by Heat *(i.e.,* by Flames, Hot Metal etc.)
- For very minor burns, hold the burn in cold saturated (8%) sodium bicarbonate solution for some time, then cover with zinc oxide ointment (or, failing this, with Vaseline) and bandage to exclude air.
- For major burns, do not put an oil or ointment, but always apply the acriflavine emulsion freely and without delay. If the burn is either on the hand or on the arm, after applying the emulsion cover the burn lightly with a layer of cotton wool also soaked in the acriflavine emulsion.

[B] Scalds (by boiling water): Apply the Acriflavine.

[C] Acid Burns:
Wash immediately with cold water and then with dilute (8%) sodium bicarbonate solution. If burn is severe, wash again with water and then apply the acriflavine or picric acid treatment.

[D] Caustic Alkali Burns:

The same as for acid burns, except that after washing with water, wash with very dilute (e.g., 1%) acetic acid solution in place of sodium bicarbonate. Then continue as before.

[E] Sodium Burns:

Most frequently caused by small molten pellets flying out of heated tubes. If a small solidified pellet of sodium can be seen, remove carefully with forceps. Then wash the burn thoroughly in water, then in dilute (1%) acetic acid, and then cover with olive oil. For serious apply the acriflavine treatment.

EYE ACCIDENTS:

In all the cases, the patient should be sent to doctor for examination. If an accident appears serious, a patient should be sent to doctor before first aid is applied. On normal accident patient should be sent to a doctor after applying first aid.

[A] Acid in Eye:

If acid is dilute, wash the eye repeatedly with 1% sodium bicarbonate solution. If the acid is concentrated, first wash the eye well with water, and then continue with the bicarbonate solution.

[B] Caustic Alkali in Eye:

Proceed just as for acid in the eye, but wash with 1% boric acid solution in place of bicarbonate solution.

[C] Glass in Eye:

Remove glass washing water in an eye bath. Soreness which may follow very minor accidents to the eye may be relieved by placing 1 drop of castor oil in the corner of the eye.

TREATMENT OF FIRES:

Laboratories should be equipped with a sufficient number of fire proof blankets, so that a blanket are available at any point of the laboratory at a few seconds notice. Each blanket should be kept in a clearly labelled box, the lid of which is closed by its own weight and not by any mechanical fastening, which might delay removal of the blanket. The box itself should be kept in some open and unencumbered position in the laboratory.

The blanket when required should be at once wrapped firmly around the person whose clothes are on fire, the person then placed in apron position on the floor with the ignited portion upwards, and the blanket pressed firmly over the ignited clothes until the fire is extinguished

Most of the available fire-extinguishers are unsuitable for chemical laboratory use. Those which give a stream of water are useless for extinguishing burning ether, benzene, petrol, etc. and exceedingly dangerous if metallic sodium or potassium are present. Those which give a vigorous and fine stream of carbon tetrachloride although of greater use than the water type, frequently serve merely to fling the burning material (and particularly burning solvents) along the surface of the bench without extinguishing the fire, the area of which is thus an actually increased.

The following methods should therefore generally be used:

(1) Sand, buckets of dry sand for fire-extinguishing should be available in the laboratory and should be strictly reserved for this purpose, and not encumbered with sand-baths, wastepaper, etc. Most fires on the bench may be quickly smothered by the ample use of sand. Sand once used for this purpose should always be thrown away afterwards, and nor returned to the buckets, as it may contain appreciable quantities of inflammable, non-volatile materials (e.g. nitrobenzene), and be dangerous if used second time.

(2) If a liquid which is being heated in a beaker or a conical flask catches fire, it is frequently sufficient to turn off the gas (or source of heating) below and then at once to stretch a clean duster tightly over the mouth of the vessel. The fire quickly dried out from lack of air, and the solution is recovered unharmed.

(3) Students should bear in mind that the majority of bench fires arise from one of three cause, all of which result from careless manipulation by the student himself. These causes are :

- The cracking of glass vessels which are being heated with first containing inflammable liquids. This cracking may occasionally be due to faulty apparatus, but is almost invariably caused by an unsuitable method of heating, the latter furthermore being often hastily applied.
- The addition of unglazed porcelain to a heated liquid which is "bumping" badly- with the result that the previously superheated liquid froths over and catches fire. Porcelain should never be added to a "bumping" liquid until the latter has been allowed to cool for a few minutes and therefore has fallen in temperature below its boiling. The most dangerous solvent in the laboratory is carbon disulphide, the flash-point of which is so low that its vapour is ignited. Carbon disulphide should, therefore never be used in the laboratory unless an adequate as a solvent cannot be found. Probably the next most dangerous liquid for general manipulation is ether, which, however has frequently to be employed. If the precautions are always followed, the manipulation of ether should however be quite safe.

EXPLOSION:

Gaseous explosions also rank among those accidents which are almost invaribly due to careless work. They are usually caused by:

- Faulty condensation of a heavy inflamable vapour, such as ether. The precautions mentioned in the previous paragraph if observed will prevent this occurrence.
- Igniting an inflamable gas before all air has been removed from the containing vessel. Whenever an inflamable gas is collected, a sample in a small test-tube should first be ignited at a safe distance from the main experiment. If it burns quietly, without any sign of even a gentle explosion, the main body of the gas can be safely ignite, although even then, this should be done with the smallest volume of gas suitable for the purpose concerned.
- Experiments in which metallic sodium has been used (e.g. the preparation of ethyl acetoacetate), and in which the product has subsequently to be treated with water, great care should be taken to ensure that sodium remains when the water is added.

SPECIAL CAUTION:

Safety goggles should always be worn over the eyes when carrying out potentially dangerous operations, eg. vacuum distillations, distillation of large volumes of inflammable liquids and experiments employing large quantities of metallic sodium.

LABORATORY APPARATUS

Before going into the laboratory or working with chemicals students must be aware with the glasswares which are in use in laboratory practice. Some common glasswares used in the laboratory are burette, pipette, conical Flask, test tubes, boiling tubes, beakers, round-bottommed flask (R.B. flask), flat-bottommed flask (Erlenmeyer flask), funnel, Hirsch funnel, Buchner funnel, separatory funnel, distillation set, condenser, distilling column, simple bent adapter, Claisen distilling head and vacuum adapter

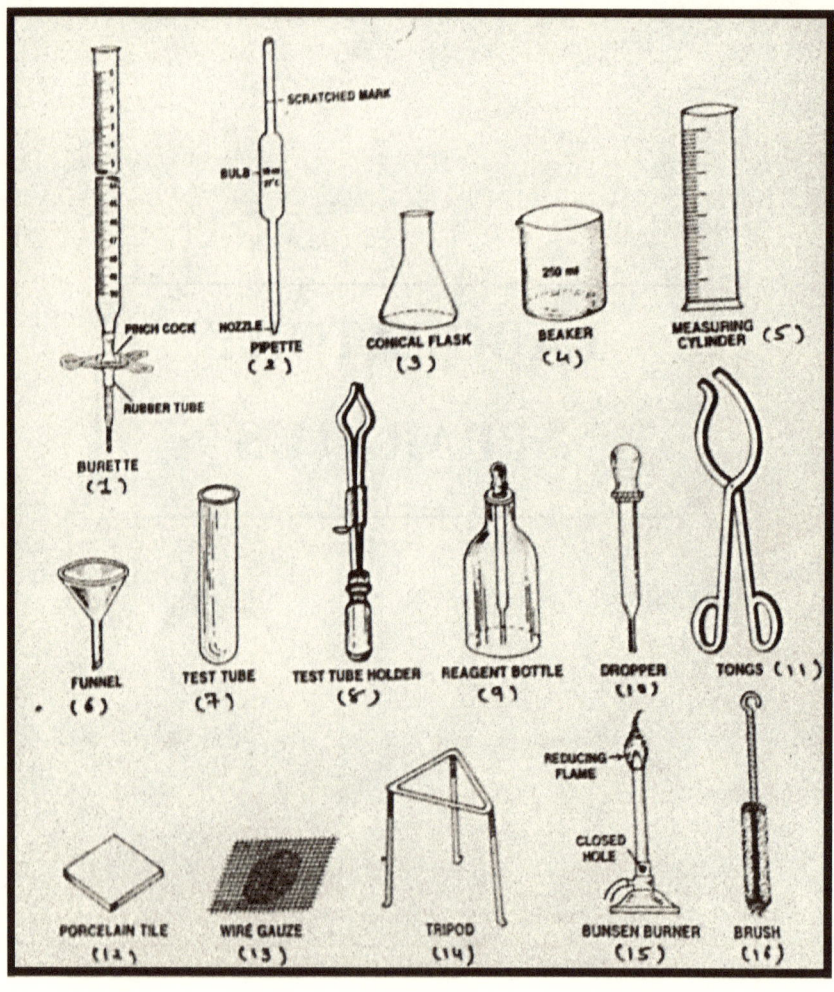

(1) Burette:

It is a long glass tube graduated from either 0.0 ml to 25.0 ml or 0.0 ml to 50.0 ml. Each 'ml' being divided further into ten equal parts i.e., 0.1 ml intervals. At the bottom, it is provided with a glass tap or with a jet tube connected by means of a rubber tubing attached with a pinch-cock. Exact volumes of liquid can be delivered by means of the burette.

(2) Pipette:

It is a glass apparatus by means of which a definite volume of a liquid can be delivered from one container to another. It is a long narrow tube provided with a bulb in the middle and drawn into a nozzle at the lower end. There is a mark scratched on the upper stem which indicates the capacity of each pipette. Usually, in practice pipettes are of the capacities 5 ml, 10 ml, 25 ml etc.

(3) Conical flask:

It is a glass vessel having a wide neck, belly and a flat bottom. Generally conical flasks of 10, 50, 100 or 250 ml capacity are used. Conical flasks are commonly used for taking solution for titration and for storing as well as heating the solution.

(4) Beaker:

Beakers with different capacity are used for storing as well as heating the solution.

(5) Measuring cylinder:

Measuring cylinders of different sizes are available to prepare standard solutions of known volume. Generally measuring cylinders of 100 ml, 250 ml, 500 ml or 1000 ml capacity are used.

(6) Funnel:

A funnel is used to transfer a liquid or solution from one container to another. It is also used for filling the burette with a solution. Keeping properly folded filter paper in a funnel, it is also used for filtration of solution or precipitate.

(7) Test tube:

It is a glass tube closed at one end. It is mainly used for performing tests in a qualitative analysis.

(8) Test tube holder:

To heat the solution in a test tube at higher temperature, it is holded by a test tube holder.

(9) Reagent bottle:

In the laboratory glass bottles are used for storing different reagents.

(10) Dropper:

Dropper is used to transfer reagent from reagent bottle to test tube.

(11) Tong:

Tong is used for holding the hot apparatus like test-tube, beaker etc.

(12) Porcelain tile:

A white porcelain tile is placed on the base of the burette stand under a burette, so that during volumetric analysis, the change in color at end point can be observed easily.

(13) Wire gauze:

To heat the solution, glass apparatus like beaker, conical flask etc. containing solution are placed on a wire gauze.

Central portion of a wire gauze is covered with asbestos in order to prevent the direct contact of the flame to the apparatus and to avoid the possibility of the breakage of glass apparatus.

(14) Tripod:

This is a three legged metal apparatus. A crucible or a beaker is placed on a clay pipe triangle or a wire gauze kept on the tripod stand for heating.

(15) Bunsen burner:

In laboratory, bunsen burner is used for heating a substance or a solution. It consists of a brass tube covered with a small metal ring at the lower end of the base, which is free to rotate. The tube and the ring has a hole on both the sides. The ring is rotated in such a way so that the hole of the internal tube remains closed or open, so as to get reducing flame or oxidizing flame.

(16) Brush:

A brush is used for cleaning test tubes.

(17) Flasks

See Fig. (2a-2e). Type (a) is a flat-bottommed flask with thin long neck of capacities between 15 ml to 2 L. The larger sizes have tooled ring neck to increase its mechanical strength. Type (b) is a round-bottomed flask with capacities 50 ml to 2 L. Type (c) is a short necked boiling flask of capacities between 50 ml to 1 L. Type (d) is a pear shaped flask (5 ml-100 ml capacity) with interchangeable ground glass joints for semi-micro work. The joint sizes are in the range 10/19 to 24/19. Type (e) is a quick fit short necked R.B. flask in the capacity range 50 ml to 2 L. The interchangeable quick fit joints of sizes 14/23, 19/26, and 24/29 are suitable for macro-scale experiments.

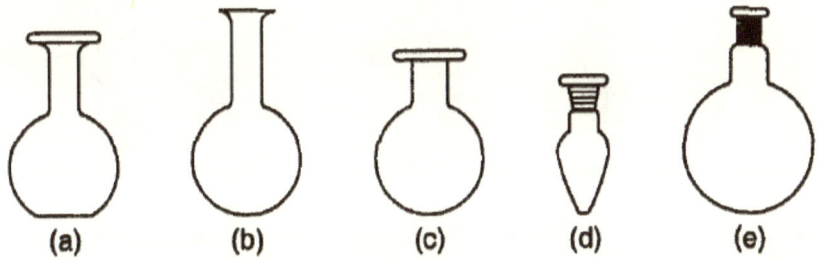

(a)   (b)   (c)   (d)   (e)

Figure 2 Different types of Flasks

(18) Two and Three Necked Flasks

If in any preparation during stirring process, it is necessary to add some reagent in the substrate drop wise or stirring is to be done in an inert atmosphere, a two or three necked flask is used as the reaction vessel (Fig. 3). The central neck is fitted with a mercury seal stirrer. The other necks are used for fitting of dropping funnel, reflux condenser or a gas inlet etc., as required.

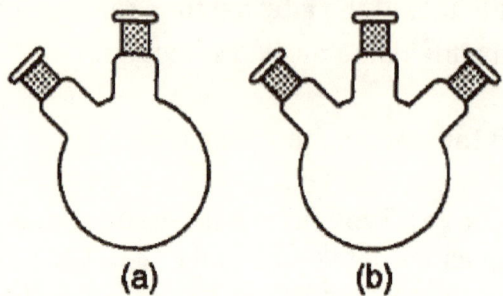

Figure 3 Two and Three neck flasks

(19) Ground Glass Stoppers and Adapters

These are available of all sizes but those with a flat head are preferred as they are easy to handle and keep clean Fig. (4a-c). A reduction adapter Fig. (3b) and an extension adapter Fig. (4c) are required if the joints sizes of various parts are not compatible.

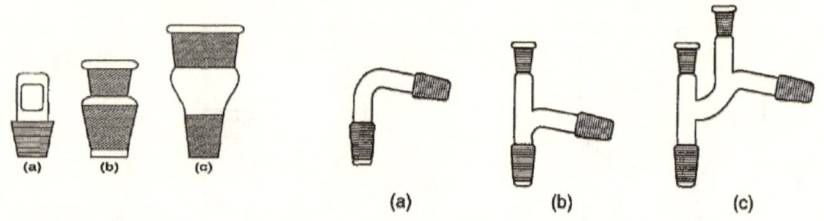

Figure 4 Stoppers   **Figure 5 distillation heads and adapters**

(20) Distillation Heads (Still Heads)

Different types of distillation heads are shown in Fig. 5(a-c). Type (a) is used in simple distillation for the removal of solvent. It is called a bend or "Knee tube". Type (b) is a simple distillation head in which a thermometer can be fitted. Type (c) is a Claisen-distillation head. The left-hand socket of this still head accommodates the capillary tube for use in distillation under vacuum and the right-hand socket accommodate thermometer.

(21) Condensers

Several types of condensers are widely used Fig. 6(a-c). Type (a) is a Liebig's or West condenser. Type (b) is a Davies condenser and type (c) is a double coil type condenser. These are highly efficient double surface condensers. These condensers are employed for both reflux and for downward distillation.

(22) Receiver Adapters

Receiver adapters are also called connectors and are attached to the end of condensers used in a distillation assembly. Different types of adapters are given in Fig. 7 (a to c).

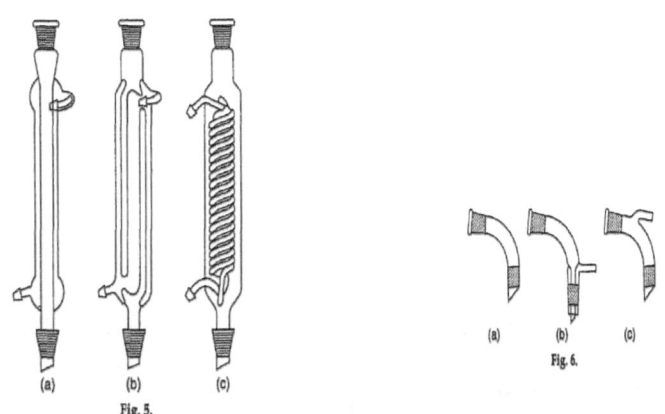

Figure 6 Condensers   Figure 7 Receiver adapter

(23) Separatory or Dropping Funnel

Several types of separatory funnels are widely used Fig. 8(a-e). Type (a) is called pear shaped, type (b) is called conical and type (c) is called cylindrical separatory funnels. The separatory funnels fitted with Teflon stopcocks are preferred as the fear of tap seisure and use of a lubricant is not required. Type (d) and (e) are quick-fit dropping funnels.

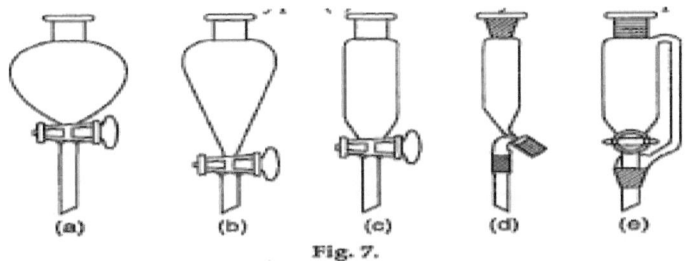

Fig. 7.

Figure 8 Dropping Funnel

Volumetric Analysis

❀ VOLUMETRIC ANALYSIS (PART – I) ❀

Exp. 1. To prepare solution of acids and bases with definite concentration.

Exp. 2. To prepare a solution by dissolving 'x' gms $NaHCO_3$ /Na_2CO_3 in 100 ml solution and determine its concentration in terms of normality and molarity using the given 0.1 M HCl solution.

Exp. 3. To determine the normality, molarity and gms/litre of NaOH and HCl using 0.05 M Na_2CO_3 solution.

Exp. 4. To determine the molarity, g/litre and normality of each component in a given mixture of $NaHCO_3$ and Na_2CO_3 using 0.1 M HCl solution.

Exp. 5. To determine the molarity, g/litre and normality of each component in a mixture of $H_2C_2O_4.2H_2O$ and H_2SO_4 using 0.02 M $KMnO_4$ and 0.1 M NaOH solution.

Exp. 6. To determine the molarity, g/litre and normality of each component in a mixture of $H_2C_2O_4.2H_2O$ and $K_2C_2O_4.H_2O$ using 0.1 M NaOH and 0.02 M $KMnO_4$ solution.

Exp. 7. To determine the molarity, g/litre and normality of $KMnO_4$ and $FeSO_4.7H_2O$ solution using 0.05 M $H_2C_2O_4.2H_2O$ solution.

Exp. 8. To determine the molarity, g/litre and normality of $FeSO_4(NH_4)_2SO_4.6H_2O$ and $K_2Cr_2O_7$ solutions using 0.02 M $KMnO_4$ solution.

❀ **INORGANIC VOLUMETRIC ANALYSIS (PART – II)** ❀

Exp. 9. Estimation of the amount of Cu^{2+} in the given $CuCl_2.2H_2O$ solution using 0.01 M EDTA solution.

Exp. 10. Estimation of the amount of Ni^{2+} in the given $NiSO_4.7H_2O$ solution using 0.01 M EDTA solution.

Exp. 11. Estimation of the amount of Zn^{2+} in the given $ZnCl_2$ solution using 0.01 M EDTA solution.

Exp. 12. Estimation of total hardness of water by EDTA.

Exp. 13. Determination of acetic acid in commercial vinegar using 0.1 M NaOH.

Exp. 14. Determination of alkali in antacid using 0.1 M HCl.

Exp. 15. Estimation of ferric (Fe^{+3}) by dichromate method (Internal indicator method).

Titration

The word titration comes from the Latin word "Titulus", which means inscription or title. The French word title means rank. Therefore, Titration means the determination of concentration or rank of a solution.

The standard solution is usually added from a graduated vessel called a burette. The process of adding standard solution until the reaction is just complete is termed as titration.

All chemical reactions cannot be considered as titrations. A reaction can serve as a basis of a titration procedure only if the following conditions are satisfied:

1. The reaction must be a fast one.
2. It must proceed stoichiometrically.
3. The change in free energy (ΔG) during the reaction must be sufficiently large for spontaneity of the reaction.
4. There should be a way to detect the completion of the reaction.

End point and Equivalent point:

For a reaction, a stage which shows the completion of a particular reaction is known as **end point**. Equivalence point is a stage in which the amount of reagent added is exactly and stoichiometrically equivalent to the amount of the reacting substance in the titrated solution. The end point is detected by some physical change produced by the solution, by itself or more usually by the addition of an auxiliary reagent known as an 'indicator'. The end point and the equivalence point may not be identical. End point is usually detected only after adding a slight excess of the titrant. In many cases, the difference between these two will fall within the experimental error.

Indicator:

It is a chemical reagent used to recognize the attainment of end point in a titration. After the reaction between the substance and the standard solution is complete, the indicator should give a clear colour change.

When a titration is carried out, the free energy change for the reaction is always negative. That is, during the initial stages of the reaction between A & B, when the titrant A is added to B the following reaction takes place.

$$A + B \rightleftharpoons C + D$$

Equilibrium constant,

$$K = \frac{a_C \cdot a_D}{a_A \cdot a_B} = \frac{[C] \cdot [D]}{[A] \cdot [B]}$$

a = Activity co-efficient.

Large values of the equilibrium constant K implies that the equilibrium concentration of A & B are very small at the equivalence point. It also indicates that the reverse reaction is negligible and the product C & D are very much more stable than the reactants A & B. Greater the value of K the larger the magnitude of the negative free energy change for the reaction between A & B. Since,

$$\text{Free Energy Change} = \Delta G = -RT \ln K$$

Where,
R = Universal gas Constant = 8.314 $JK^{-1}mol^{-1}$
T = Absolute Temperature.

The reaction of the concentration of A & B leads to the reduction of the total free energy change. If the concentrations of A & B are too low the magnitude of the total free energy change becomes so small and the use of the reaction for titration will not be feasible.

Expressions of Concentration of Solutions:

The concentration or strength of solution means the amount of solute present in a given amount of the solution. The concentration may be expressed in physical or chemical units.

1. Normality (N):

It is defined as number of gram equivalents of the solute present in 1 litre (1000 mL.) of the solution. If W g of solute of equivalent weight E is present in V mL of the solution, the normality of the solution is given by:

$$\text{Normality} = \frac{W \times 1000}{E \times V}$$

2. Molarity (M):

It is defined as the number of moles of the solute present in 1 litre (or 1000 mL) of the solution. A one molar solution contains 1 mole of the solute dissolved in 1 litre of the solution.

3. Molality (m):

It is defined as the number of moles of solute dissolved in 1000 g of the solvent. One molal solution contains one mole of the solute dissolved in 1000 g of the solvent.

Normal solution:

A solution containing one gram equivalent weight of the solute dissolved per litre is called a normal solution; e.g. when 40 g of NaOH is present in one litre of NaOH solution, the solution is known as normal (N) solution of NaOH. Similarly, a solution containing a fraction of gram equivalent weight of the solute dissolved per litre is known as subnormal solution. For example, a solution of NaOH containing 20 g (1/2 of g eq. wt.) of NaOH dissolved per litre is a sub-normal solution. It is written as N/2 or 0.5 N solution.

Formulae used in solving numerical problems on volumetric analysis;

1. Strength of solution = Amount of substance in g litre^{-1}.

2. Strength of solution = Amount of substance in g moles litre^{-1}.

3. Strength of solution = Normality × Eq. wt. of the solute = molarity × Mol. wt. of solute.

4. Molarity = Moles of solute/Volume in litre.

5. Number of moles = Wt.in g/Mol. wt = M × V (initial) = Volume in litres/22.4 at NTP (only for gases).

6. Number of milli moles = Wt. in g × 1000/mol. wt. = Molarity × Volume in mL.

7. Number of equivalents = Wt. in g/Eq. wt = x × No. of moles × Normality × Volume in litre (Where x = Mol. wt/Eq. wt).

8. Number of milliequivalents (meq.) = Wt. in g × 1000 / Eq. wt = normality × volume in mL.

9. Normality = x × No. of millimoles (Where x = valency or change in oxi. number).

10. Normality formula, $N_1V_1 = N_2V_2$, (Where N_1, $N_2 \rightarrow$ Normality of titrant and titrate respectively, V_1, $V_2 \rightarrow$ Volume of titrant and titrate respectively).

11. % by weight = Wt. of solvent/Wt. of solution × 100.

A solution is a homogeneous mixture of two or more components, the composition of which may be changed. The substance which is present in smaller proportion is called the solute, while the substance present in large proportion is called the solvent.

Volumetric Analysis:

It involves the estimation of a substance in solution by neutralization, precipitation, oxidation or reduction by means of another solution of accurately known strength. This solution is known as standard solution.

Volumetric analysis depends on measurements of the volumes of solutions of the interacting substances. A measured volume of the solution of a substance A is allowed to react completely with the solution of definite strength of another substance B. The volume of B is noted. Thus we know the volume of the solutions A and B used in the reaction and the strength of solution B; so the strength of the other solution A is obtained. The amount (or concentration) of the dissolved substance in volumetric analysis is usually expressed in terms of normality. The weight in grams of the substance per litre of the solution is related to normality of the solution as,

Weight of the substance (g per litre) = Normality × gram equivalent weight of the substance.

Conditions of Volumetric Analysis:

(i)   The reaction between the titrant and titrate must be expressed.
(ii)   The reaction should be practically instantaneous.
(iii) There must be a marked change in some physical or chemical property of the solution at the end point.
(iv)   An indicator should be available which should sharply define the end point.

Different methods to determine the endpoint include:

• pH indicator:

A pH indicator is a substance that it changes its colour in response to a chemical change. An acid-base indicator changes its colour depending on the pH (e.g., phenolphthalein). Redox indicators are also frequently used. A drop of indicator solution is added to the titration at the start; at the endpoint has been reached the colour changes.

• Colour change:

In some reactions, the solution changes colour without any added indicator. This is often seen in redox titrations, for instance, when the different oxidation states of the product and reactant produce different colours.

• Precipitation:

In this type of titration the strength of a solution is determined by its complete precipitation with a standard solution of another substance.

eg:

$$AgNO_3 + NaCl \rightarrow AgCl + NaNO_3$$

Acid base titration:

The chemical reaction involved in acid-base titration is known as neutralisation reaction. It involves the combination of H_3O^+ ions with OH- ions to form water. In acid-base titrations, solutions of alkali are titrated against standard acid solutions. The estimation of an alkali solution using a standard acid solution is called *acidimetry*. Similarly, the estimation of an acid solution using a standard alkali solution is called *alkalimetry*.

The Theory of Acid–Base Indicators:

Ostwald, developed a theory of acid base indicators which gives an explanation for the colour change with change in pH. According to this theory, a hydrogen ion indicator is a weak organic acid or base. The undissociated molecule will have one colour and the ion formed by its dissociation will have a different colour.

Let the indicator be a weak organic acid of formulae HIn. It has dissociated into H^+ and In^-. The unionized molecule has one colour, say colour (1), while the ion, In^- has a different colour, say colour (2). Since HIn and In^- have different colours, the actual colour of the indicator will dependent upon the hydrogen ion concentration $[H^+]$. When the solution is acidic, that is the H^+ ions present in excess, the indicator will show predominantly colour (1). On other hand, when the solution is alkaline, that is, when OH⁻ ions present in excess, the H⁺ ions furnished by the indicator will be taken out to form undissociated water. Therefore there will be larger concentration of the ions, In⁻ thus the indicator will show predominantly colour (2).

Some indicators can be used to determine pH because of their colour changes somewhere along the change in pH range. Some common indicators and their respective colour changes are given below.

Indicator	Colour on Acidic Side	Range of Colour Change	Colour on Basic Side
Methyl Violet	Yellow	0.0 - 1.6	Violet
Bromophenol Blue	Yellow	3.0 - 4.6	Blue
Methyl Orange	Red	3.1 - 4.4	Yellow
Methyl Red	Red	4.4 - 6.2	Yellow
Litmus	Red	5.0 - 8.0	Blue
Bromothymol Blue	Yellow	6.0 - 7.6	Blue
Phenolphthalein	Colorless	8.3 - 10.0	Pink
Alizarin Yellow	Yellow	10.1 - 12.0	Red

i.e., at pH value below 5, litmus is red; above 8 it is blue. Between these values, it is a mixture of two colours.

Indicators Used for Various Titrations:

1. Strong Acid against a Strong Base:

Let us consider the titration of HCl (strong acid) and NaOH (strong base). The pH values of different stages of titration shows that, at first the pH changes very slowly and rise to only about 4. Further addition of such a small amount as 0.01 mL of the alkali raises the pH value by about 3 units to pH 7. Now the acid is completely neutralized. Further of about 0.01 mL of 0.1 M NaOH will amount to adding hydrogen ions and the pH value will jump to about 9. Thus, near the end point, there is a rapid increase of pH from about 4 to 9.

An indicator is suitable only if it undergoes a change of colour at the pH near the end point. Thus the indicators like **methyl orange, methyl red and phenolphthalein** can show the colour change in the pH range of 4 to 10. Thus, in strong acid-strong base titrations, any one of the above indicators can be used.

2. Weak Acid against Strong Base:

Let us consider the titration of CH_3COOH (weak acid) against NaOH (strong base). The titration shows the end point between pH 8 to 10. This is due to the hydrolysis of sodium acetate formed. Hence **phenolphthalein** is a suitable indicator as its pH range is 8-9.8. However, methyl orange is not suitable as its pH range is 3.1 to 4.5.

3. Strong Acid against Weak Base:

Let us consider the titration of HCl (strong acid) against NH_4OH (weak base). Due to the hydrolysis of the salt, NH_4Cl is formed during the reaction, the pH lies in the acid range. Thus, the pH at end point lies in the range of 6 to 4. Thus **methyl orange** is a suitable indicator while phenolphthalein is not suitable.

Strong Acids	Strong Bases	Weak Acids	WeakBases
HCl	NaOH	Acetic acid	Ammonia
HNO_3	KOH	Hydrocyanic acid	Magnesium hydroxide
HBr	-	HF	Pyridine
H_2SO_4	-	Oxalic acid	Sodium carbonate
HI	-	Ethanoic acid	Potassium carbonate

Precipitation Titration:

A titrimetric method based on the formation of a slightly soluble precipitate is called a precipitation titration. The most important precipitation process in titrimetric analysis utilizes silver nitrate as the reagent (Argentimetric process).

$$Ag^+_{(aq)} + Cl^-_{(aq)} \rightleftharpoons AgCl_{(s)}$$

Many methods are utilized in determining end points of these reactions, but the most important method, the formation of a coloured precipitate will be considered here.

1.   In the titration of a neutral solution of chloride ions with silver nitrate, a small quantity of potassium chromate solution is added to serve as the indicator. At the end point the chromate ions combine with silver ions to form the sparingly soluble brick-red silver chromate. This is a case of fractional precipitation, the two sparingly soluble salts being AgCl (Ksp = 1.2 x 10^{-10}) and Ag_2CrO_4 (Ksp = $1.7x10^{-12}$).

AgCl is the less soluble salt and initially chloride concentration is high, hence AgCl will be precipitated. Once the chloride ions are over and with the addition of small excess of silver nitrate solution brick red coloured silver chromate becomes visible. The titration should be carried out in neutral solution or in very faintly alkaline solution. i.e. within the pH range 6.5-9.

In acid solutions following reaction occurs.

$$2CrO_4^{2-} + 2H^+ \rightleftharpoons 2HCrO_4^-$$

$$\updownarrow$$

$$2Cr_2O_7^{2-} + H_2O$$

Consequently the chromate ions concentration is reduced and the solubility product of silver chromate may not be exceeded. In markedly alkaline solution, silver hydroxide (Ksp = 2.3 x 10$^{8)}$) might be precipitated.

2. The titration can be carried out with dichlorofluorescein as the indicator. Dichlorofluorescein is an example of an adsorption indicator. Adsorption indicators have the interesting property of changing colour when they stick (adsorb) to the surface of a precipitate. During the titration the dichlorofluorescein molecules exist as negatively charged ions (anions) in solution. As the AgCl precipitate is forming, the excess Cl$^-$ ions in the solution form a layer of negative charge on the precipitate surface. As the equivalence point is reached and passed, the excess Cl$^-$ ions on the precipitate surface are replaced by excess Ag+ ions, giving the surface a positive charge. The negatively charged indicator will be attracted to the positively charged precipitate surface where it absorbs and changes colour. The suspended precipitate will have a pink tinge because of some premature displacement of chloride ion by the dichlorofluorescein ion. When the pink colour starts to persist for slightly longer periods of time, the drip rate is lowered. The end point is reached when the entire solution turns pink. It is important that the AgCl precipitate be prevented from coagulation during the titration. For this reason a small amount of dextrin is added to the solution.

Complexometric Titration:

This type of titration depends upon the combination of ions (other than H$^+$ and OH$^-$) to form a soluble ion or compound as in the titration of a solution of a cyanide with AgNO$_3$.

Principle of Complexometric Titration:

Complexometric titrations are particularly useful for determination of a mixture of different metal ions in solution. Ethylene diamine tetra acetic acid (EDTA), is very important reagent for complex formation titrations. EDTA has been assigned the formula II in preference to I since it has been obtained from measurements of the dissociation constants that two hydrogen atoms are probably held in the form of zwitter ions.

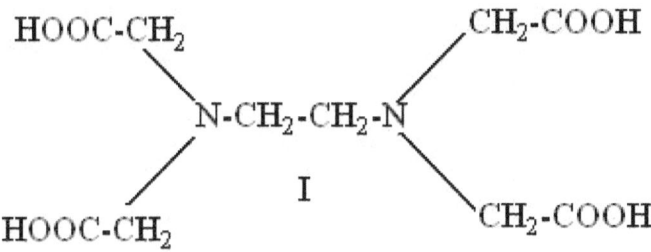

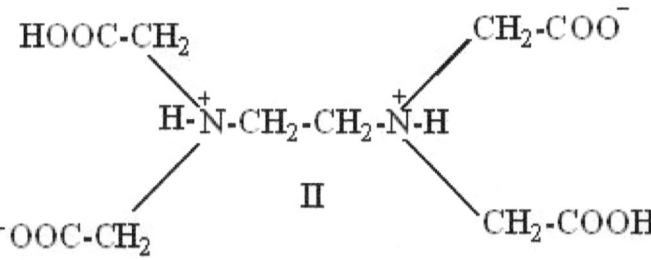

EDTA behaves as a dicarboxylic acid with two strongly acidic groups. For simplicity EDTA may be given the formula H_4Y, the disodium salt is therefore Na_2H_2Y and it has the complex forming ion H_2Y^{2-} in aqueous solution. The reactions with cations may be represented as;

$$M^{2+} + H_2Y^{2-} \rightarrow MY^{2-} + 2H^+$$
$$M^{3+} + H_2Y^{2-} \rightarrow MY^- + 2H^+$$
$$M^{4+} + H_2Y^{2-} \rightarrow MY + 2H^+$$

One gram ion of the complex-forming ion H_2Y^{2-} reacts in all cases with one gram ion of the metal. EDTA forms complexes with metal ions in basic solutions. In acid-base titrations the end point is detected by a pH sensitive indicator. In the EDTA titration metal ion indicator is used to detect changes of pM. It is the negative logarithm of the free metal ion concentration, i.e., $pM = - \log [M^{2+}]$. Metal ion complexes form complexes with specific metal ions. These differ in colour from the free indicator and a sudden colour change occurs at the end point. End point can be detected usually with an indicator or instrumentally by potentiometric or conductometric (electrometric) method.

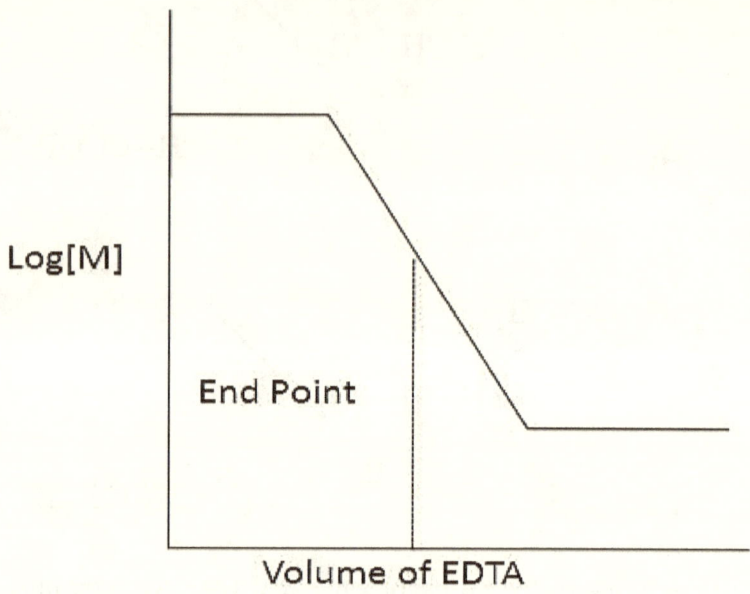

Volume of EDTA

There are three factors that are important in determining the magnitude of break in titration curve at end point.

• **The stability of complex formed:** The greater the stability constant for complex formed, larger the charge in free metal concentration (pM) at equivalent point and more clear would be the end point.

• **The number of steps involved in complex formation:** Fewer the number of steps required in the formation of the complex, greater would be the break in titration curve at equivalent point and clearer would be the end point.

• **Effect of pH:** During a complexometric titration, the pH must be constant by use of a buffer solution. Control of pH is important since the H^+ ion plays an important role in chelation. Most ligands are basic and bind to H^+ ions throughout a wide range of pH. Some of these H^+ ions are frequently displaced from the ligands (chelating agents) by the metal during chelate formation.

• Equation below shows complexation between metal ion and H^+ ion for ligand:

$$M_2^+ + H_2\text{-EDTA} \rightarrow M\text{-EDTA} + 2H^+$$

Thus, stability of metal complex is pH dependent. Lower the pH of the solution, lesser would be the stability of complex (because more H^+ ions are available to compete with the metal ions for ligand). Only metals that form very stable complexes can be titrated in acidic solution, and metals forming weak complexes can only be effectively titrated in alkaline solution.

Mechanism of action of indicator:

During an EDTA titration two complexes are formed: i) M-EDTA complex and ii) M-indicator complex.

$$M\text{-In} + EDTA \rightarrow M\text{-EDTA} + In$$

Eriochrome black T is a metal ion indicator. In the pH range 7 - 11 the dye itself has a blue colour. In this pH range addition of metallic salts produces a brilliant change in colour from blue to red.

$$M^{2+} \quad + \quad HIn^{2-} \quad \rightarrow \quad MIn^- \quad + \quad H^+$$
$$\text{(Blue)} \qquad\qquad\qquad\qquad \text{(Red)}$$

This colour change can be obtained with the metal ions. As the EDTA solution is added, the concentration of the metal ion in the solution decreases due to the formation of metal-EDTA complex. At the end point no more free metal ions are present in the solution. At this stage, the free indicator is liberated and hence the colour changes from red to blue.

Indicators used in complexometric titrations are as follows:

Sr. No.	Name of indicator	Colour change	pH range	Metals detected
1	Mordant black II Eriochrome black T Solochrome black T	Red to Blue	6-7	Ca, Ba Mg, Zn, Cd, Mn, Pb, Hg
2	Murexide or Ammonium purpurate	Violet to Blue	12	Ca, Cu, Co
3	Catechol-violet	Violet to Red	8-10	Mn, Mg, Fe, Co, Pb
4	Methyl Blue	Blue to Yellow	4-5	Pb, Zn, Cd, Hg
	Thymol Blue	Blue to Grey	10-12	
5	Alizarin	Red to Yellow	4.3	Pb, Zn, Co, Mg, Cu
6	Sodium Alizarin sulphonate	Blue to Red	4	Al, Th
7	Xylenol Orange	orange yellow to purple	1-3	Bi, Th
			4-5	Pb, Zn
			5-6	Cd, Hg

Alizarin fluorine blue(alizarin complexone)

Calcone (mordant black 17)

Eriochrome Blue Black R

Calcon carboxylic acid

Catechol - violet

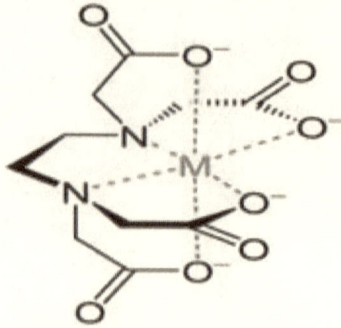

(Metal-EDTA complex)

Applications of Complexometric titration:

- Complexometric titration is widely used in the medical industry because of the micro litre size sample involved. The method is efficient in research related to the biological cell.
- Ability to titrate the amount of ions available in a living cell.
- Ability to introduce ions into a cell in case of deficiencies. Complexometric titration involves the treatment of complex ions such as magnesium, calcium, copper, iron, nickel, lead and zinc with EDTA as the complexing agent.
- Complexometric titration is an efficient method for determining the level of hardness of water.

Types of Complexometric Titration:

As mentioned earlier, EDTA is a versatile chelating titrant that has been used in innumerable complexometric determinations. The versatility of EDTA can be ascribed to the different ways in which the complexometric titration can be executed. Let us learn about different ways in which we can use EDTA titrations.

1. **Direct Titration:** It is the simplest and the most convenient method in which the standard solution of EDTA is slowly added to the metal ion solution till the end point is achieved. It is similar to simple acid-base titrations. For this method to be useful the formation constant must be large and the indicator must provide a very distinct colour change as mentioned

earlier. Further we need standardized solution of EDTA and sometimes auxiliary complexing agents may be required. Some important elements which could be determined directly by the complexometric titration are Cu, Mn, Ca, Ba, Br, Zn, Cd, Hg, Al, Sn, Pb, Bi, Cr, Mo, Fe, Co, Ni, and Pd, etc. However, the presence of other ions may cause interference and need to be suitably handled.

2. **Back Titration:** In this method, an excess of a standard solution of EDTA is added to the metal solution being determined so as to complex all the metal ions present in the solution. The excess of EDTA left after the complex formation with the metal is back titrated with a standard solution of a second metal ion. This method becomes necessary if the analyte precipitates in the absence of EDTA or reacts too slowly with EDTA, or it blocks the indicator. For example, determination of Mn is done by this method because a direct titration is not possible due to precipitation of $Mn(OH)_2$. The excess EDTA remaining after complexation, is back titrated with a standard Zn solution using Eriochrome black T as indicator. However, one has to ensure that standard metal ion should not displace the analyte ion from their EDTA complex.

3. **Replacement Titration:** When direct or back titrations do not give sharp endpoints or when there is no suitable indicator for the analyte the metal may be determined by this method. The metal to be analyzed is added to a metal-EDTA complex. The analyte ion (with higher K_f) displaces EDTA from the metal and the metal is subsequently titrated with standard EDTA. For example, in the determination of Mn an excess of Mg EDTA chelate is added to Mn solution. The Mn ions quantitatively displace Mg from Mg-EDTA solution because Mn forms a more stable complex with EDTA. The free Mg metal is then directly titrated with a standard solution of EDTA using Eriochrome black T indicator. Ca, Pb and Hg may also be determined by this method.

4. **Indirect Titration:** Certain anions that form precipitate with metal cations and do not react with EDTA can be analyzed indirectly. The anion is first precipitated with a metal cation and the precipitate is washed and boiled with an excess of disodium EDTA solution to form the metal complex. The protons from disodium EDTA are displaced by a heavy metal and titrated with sodium alkali. Therefore, this method is also called alkalimetric titration. For example, barbiturates can be determined by this method.

Redox titration:

A reaction in which one or more electrons are lost is known as *oxidation* and a reaction in which one or more electrons are gained is known as *reduction*. Accordingly, a substance which can accept one or more electrons is known as *oxidizing agent* and a substance which can donate one or more electrons is called *reducing agent*. Titrations of this type are called *redox titrations*. Thus, redox titrations are those involving transfer of electrons from the reducing agent to the oxidizing agent.

Potassium permanganate, potassium dichromate, ceric sulphates, etc., are the common oxidizing agents used in redox titrations. Oxalic acid, Mohr's salt and arsenious oxide are reducing agents commonly used in redox titrations.

✰ APPARATUS USED IN VOLUMETRIC ANALYSIS ✰

The apparatus used for measuring accurately the volume of the solutions in the volumetric analysis can be classified into two groups. They are:

(1) Graduated Apparatus: Burette, pipette, etc.

(2) General Apparatus: Conical flasks, porcelain tiles, glass rod, funnel, wash bottle, etc.

✿ Precautions to be taken while using a burette :

(1) The burette is washed thoroughly with ordinary water (tap water) at first and then with distilled water. It is finally rinsed with the solution (2-3 ml) for which it is to be filled.

(2) It should be clamped vertically in the burette stand at a convenient height.

(3) The stop-cock or pinch-cock of burette should be properly greased to avoid leaking of the solution.

(4) A small funnel can be used to fill the solution in order to avoid wastage. The funnel should be removed at once after filling the solution in the burette.

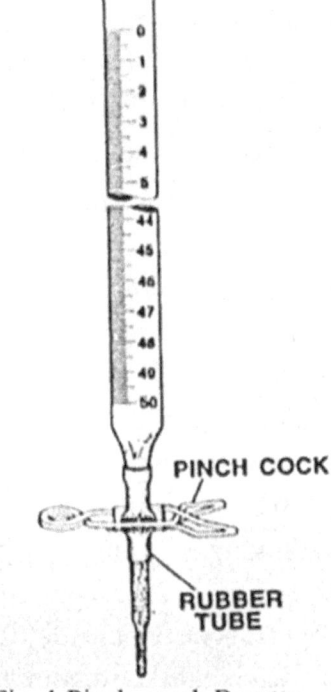

(5) Do not forget to remove the air bubbles from the nozzle or pinch-cock. After filling the burette upto the mouth with the solution, the air bubbles should be removed by rapidly running down a little of the solution through the tip or the burette by turning the tip in the upward direction [Refer fig. 3].

Fig. 1 Pinch-cocrk Burette

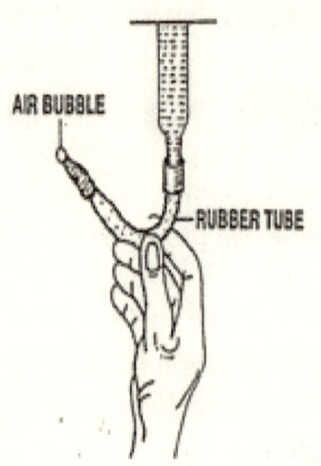

Fig. 2 Removing air bubble from the jet of the burette

(6)   While taking the readings, take care that there is no drop of solution hanging at the tip of the nozzle. If there is any, remove it by touching it with a piece of paper.

(7)   While recording a burette reading, keep your eye in level with the level of the solution in the burette. A wrong reading will be recorded if you keep your eye too high or too low as shown in Fig. 3 (An antiparallax card can also be used to take correct readings.)

(8) Read the lower meniscus in the case of colorless and transparent liquids, and read the upper meniscus in the case of deeply colored liquids, e.g. $KMnO_4$ solution.

(9) During the process of titration, the burette solution should be allowed to run slowly in order to avoid experimental error. The solution in the conical flask is stirred with one hand, so that the two solutions get mixed thoroughly and react completely.

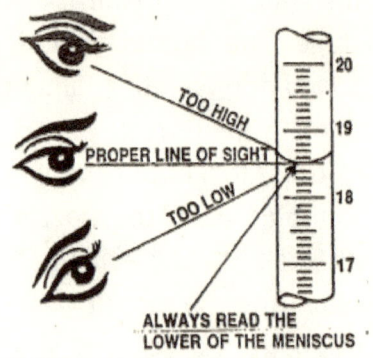

Fig. 3 Proper eye level for reading the volume

(10)   It is not necessary to coincide the level of the solution with zero while taking remaining readings.

(11)   On completion of the experiment the solution of the burette should be removed immediately and the burette should be thoroughly rinsed with water.

❀ **Precautions to be taken while using a pipette :**

(1)   The pipette should be washed thoroughly with distilled water.

(2)   It should be rinsed with the solution to be delivered.

(3)   While sucking the solution, the tip of the pipette should be well below the level of the solution in the container [Refer Fig. 5(A)]

(4)   Suck the solution in the pipette till the level of the solution rises above the scratched mark on the pipette, and press tightly the upper end of the pipette with the first finger [Refer Fig. 5 (B)].

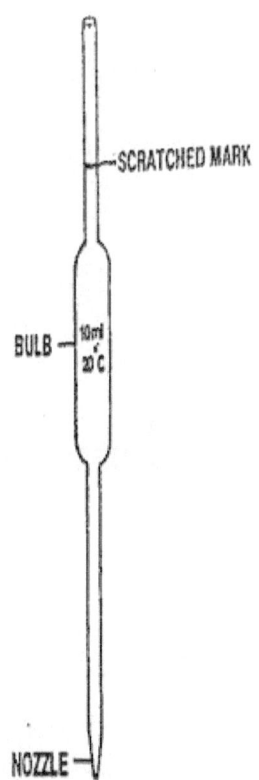

(5)   Now remove the pipette from the solution and hold it vertically. Keep the eye in level with the mark on the pipette. Then release the pressure of the finger slightly to allow the level to sink slowly.

(6)   When the level of the solution just coincides with the scratched mark on the pipette, stop the flow of the solution by pressing the finger tightly [Fig. 5 (C)].

(7)   Lift the finger from the mouth of the pipette and allow this measured volume of the solution to run into the conical flask [Fig. 5 (D).

Fig 4 Pipette

(8)   When the solution has drained off, the pipette should simply touch the side of the conical flask [Fig. 5 (E)].

(9)   The last drop, if it sticks to the jet, should not be blown out forcibly.

(10) Do not suck solutions of strong acids of strong bases in the pipette, otherwise they go into the mouth.

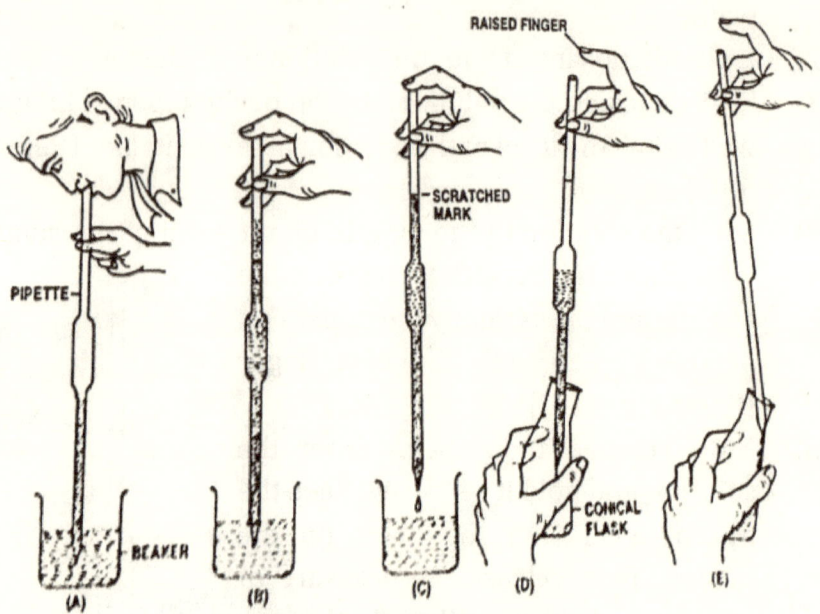

Fig. 5 Correct method of using pippete

✿ **Preparation of the solution required for the Practical:**
✓ **0.01 M E.D.T.A.**

Dissolve 3.72 gm disodium salt of EDTA in 250 ml water and make up the solution to 1000 ml with measuring flask to prepare 0.01 M EDTA solution.

✓ **6 N Ammonium Hydroxide**

Your stock solution of ammonium hydroxide is calculated to be 14.534 N based on a density of 0.9 g/mL, a formula weight of 35.05 g/mol, and a concentration of 56.6% w/w. To make 6 N solution, take 412.839 mL of your stock and adjust the final volume of solution to 1000 mL with deionizer water.

✓ **Fast Sulphon Black F**

Sodium salt of 1-hydroxy-8-(2-hydroxynaphthylazo)-2-(solphonaphtylazo)-3, 6-disulphinic acid (V). The indicator solution is 0.5% solution in water. Specific color change for copper is from magenta to pale blue to bright green.

✓ **0.1 M NaOH**

Dissolve 4 gm sodium hydroxide in 250 ml water; allow it to cool down and make up the solution to 1000 ml with measuring flask to prepare 0.1 M NaOH solution.

✓ **0.1 M HCl**

Your stock solution of hydrochloric acid is calculated to be 12.178 M based on a density of 1.2 g/mL, a formula weight of 36.46 g/mol, and a concentration of 37% w/w. To make a 0.1 M solution, slowly add 8.212 mL of your stock solution to 250 mL deionized water. Adjust the final volume of solution to 1000 mL with deionized water.

✓ **0.1 N $K_2Cr_2O_7$**

M. W. = 294.185; Eq. Wt = 49.03. Dissolve 4.90 gm $K_2Cr_2O_7$ in 250 ml water and make up the solution to 1000 ml with measuring flask to prepare 0.1 N $K_2Cr_2O_7$ solution.

✓ **Eriochrome Black T**

Sodium1-(1-hydroxy-2-naphthalyzo)-6-intro-2-naphthol-4-sulphonate (1); also known as Solochrome Black T. 0.4% w/v solution of EBT in methanol may last for about 1 to 2 months. Color change is from blue to wine red.

✓ **Phenolphthalein**

Dissolve 0.5 g of the reagent in 50 ml of alcohol and add 50 ml of water with stirring. Filter if precipitate forms. Color changes from colorless to pink.

✓ **Murexide (metal-ion EDTA)**

It is ammonium salt of purpuric acid. Suspend 0.5 g of the powdered dye stuff in water, shake thoroughly and allow to settle. The saturated supernatant is used as the indicator. The color change is towards a blue endpoint.

✓ **Diphenylamine :**

Dissolve 1 g in 100 ml conc. H_2SO_4 . Color change = blue-violet to colorless.

✓ **5% w/w $HgCl_2$**

Dissolve 5 g $HgCl_2$ in 100 ml double distilled water.

✓ **Ammonia-ammonium chloride buffer solution (pH=10)**

Dissolve 17.5 g of pure NH_4Cl (A.R.) in 152 ml of liquid NH_3 (specific gravity 0.88 to 0.9) solution and dilute the solution to 250 ml with distilled water (resulting buffer solution will have pH = 10).

Important Experimental notes: (Must read it before you start titration)

1) It is very important that your burette be clean. An easy way to clean a burette is to rinse it once or twice with deionized water and then once or twice with the solution to be used. To do this, add about 3-5 mL of solution to the burette. Hold it almost horizontally so that the solution runs down the side, but not out the top. Rotate the burette so that all the surface of glass is rinsed. Run some of the solution out the tip and pour the rest out the top. (This rinsing technique also works well for pipets.)

2) Don't clean a burette or pipet with soap unless it is absolutely necessary. Soaps are a lot easier to get in than they are to get out. If you rinse the glassware carefully with deionized water after each use, you may never need to wash it with soap.

3) You can titrate into an Erlenmeyer flask or conical flask because you can swirl the flask easily without danger of loss of the contents.

4) Always check your burette before you fill it with solution to make sure that it does not leak.

5) Always place a container such as a small beaker below the burette when you are filling it. Be sure the stopcock is closed.

6) Bubbles in the burette tip can usually be removed by tapping the side of the burette while the stopcock is open. This is best done over a sink.

7) Always use a sliding double burette clamp to fasten your burette to a ring stand. It makes it easier to get under the burette and to raise or lower it to eye level for an accurate reading.

8) Read the volume at the bottom of the meniscus. You have to get your eye at the same level as the meniscus. Depending on the colour of the calibrations it is often helpful to put either a piece of white paper or dark paper behind the scale when making a reading. Some people like to use a white card with a heavy dark line to "lineup" the meniscus with the scale.

9) When filling the burette, be careful not to get solution on the outside as this can run down and eventually fall into the solution - after picking up who knows what when passing around the clamp. Some people like to use funnels when filling the burette. If you do, be sure the funnel is very clean. It is easy to run a burette over when filling with a funnel, so be careful.

10) If your hands suddenly feel soapy during an acid-base titration, you have probably spilled some of the base solution on them. Wash your hands and any slippery feeling glassware with lots of water right away. If your hands begin to itch or burn for no apparent reason, you have probably spilled some acid on them. Wash with lots of water. Be sure to tell your instructor if you spill any acid or base solutions on your skin.

11) Always place a piece of white paper under the flask to be titrated. This will help you see the color change at the end point more clearly.

12) Don't forget to add the indicator. If you cannot remember adding indicator after starting to titrate, go ahead and add 1 or 2 drops. A little extra indicator won't hurt anything but none at all will cause your titration to fail.

13) The color of a solution can interfere with that of the indicator. If your solution is colored, run a trial to see what the solution and indicator look like at various pH values before you start to titrate. Sometimes diluting a colored solution with deionized water will make the indicator more visible.

14) It is easier for the eye to distinguish a change from colorless to colored than from colored to colorless, so if phenolphthalein is your indicator it is best to titrate the acid solution with the base (although in theory you could titrate either way).

15) While titrating, it is a good idea to wash down the walls of the flask with deionized water so that all of the titrant gets into the reaction mixture. Addition of small to moderate amounts of deionized water will not change the results of your titration.

16) Consistent when reading your burette. The same eye that makes the initial reading should also make the final reading.

Any absolute errors will then subtract out. This does not mean, however, that one lab partner must do all the titrations - it just means finish the ones you start.

17) The desired end point in a phenolphthalein titration is "A pale pink tint that persists for 30 seconds". If you are not sure if you have reached this point, record the volume of the burette, then add one more drop of base. If the solution turns dark pink or red, you were right. If not, keep titrating. As a rule of thumb, endpoints that look like pink lemonade or paler are probably satisfactory - endpoints that look like cherry kool-aid or darker will probably have to be repeated.

18) Some people like to do a quick "throw away" titration. They add the titrant very rapidly up to and a little beyond the endpoint to get a general idea of how much titrant will be required. Then they can perform titrations more quickly by adding titrant rapidly until the endpoint is near and then switch to adding drop wise.

19) If your titration seems to be taking a lot of titrant (more than 1 or 2 burettes full) you may have one of the following problems:
   (a) forgot to add indicator
   (b) titrating with wrong solution
   (c) solution being titrated is too concentrated

20) Titration is a skill that takes a lot of practice to perfect. Be patient and don't give up if you want to develop a skill that will serve you well in many chemistry labs to come.

21) It almost always takes less time to do an experiment once, slowly and carefully, than to do it as fast as you can over and over until you get it right.

Exercise for Entry Level Students
Level – 1

Experiment No. 1

Aim: To prepare solutions of acids and bases with different concentrations.

Chemical Requirements: Concentrated HC1, solid Na_2CO_3.

Procedure:

☞ **To prepare 0.1 M HC1 solution.**

Take 0.85 ml conc. HC1 solution in 100 ml measuring/volumetric flask and dilute upto the mark with distill water. Shake well.

☞ **To prepare 0.05 M Na_2CO_3 solution.**

Weigh accurately 0.53 gm of Na_2CO_3 and dissolve completely in approx. 50 ml distill water. Transfer the solution to 100 ml measuring/volumetric flask and dilute upto the mark with distill water. Shake well.

Calculation

☞ **Preparation of 0.05 M Na_2CO_3 solution**
(M.W of Na_2CO_3 = 106)

106 gm of Na_2CO_3 in 1000 ml = 1 M Na_2CO_3 solution in 1000 ml

53 gm of Na_2CO_3 = 0.5 M Na_2CO_3 solution

0.53 gm of Na_2CO_3 in 100 ml = 0.05 M Na_2CO_3 solution

☞ **Preparation 0.1 M HC1 solution (M.W of HC1 = 36.5)**

(Purity = 35-37%, Calculated purity = 36% Density=1.18 g/lit)

36.5 gm HC1 in 1 lit =1 M HC1

If purity is 100% then 36.5 gm HC1 in 1000 ml = 1 M HC1 Solution

But for 36% then 101.38 gm HC1 in 1000 ml=   1   M   HC1 Solution

Now 101.38 gm HC1 in 1000 ml

i.e. 85.91 ml HC1 in 1000 ml = 1 M HC1 Solution

0.86 ml HC1 in 100 ml   = 0.1 M HC1 Solution.

❖ **RESULT:**

(i) **In this way, we prepare 100 ml, 0.05 M Na$_2$C0$_3$ solution.**

(ii) **Similarly 100 ml, 0.1 M HC1 solution is prepared**

Experiment No. 2

Aim: To prepare a solution by dissolving 'x' gms of Na_2CO_3 in 100 ml solution and determine its concentration in terms of molarity, g/litre and normality by using   0.1 N HC1.

Requirements: 0.1 M HC1, Na_2CO_3 solution, methyl orange

Procedure:

Part I: To prepare the standard solution of 'x' M Na_2CO_3

➤ Given the 'x' grams Na_2CO_3 in clean beaker.

➤ Add approximately 50 ml distill water and try to dissolve it.

➤ Transfer the solution to a 100 ml volumetric flask and make up to 100 ml with distill water.

➤ Shake the solution well and keep aside for use in Part -II

Part II: Determination of concentration of the given Na_2CO_3 solution

➤ Wash all the apparatus with water.

➤ Rinse the burette with the given 0.1 M HC1 solution and fill the burette with the given 0.1 M HC1 solution

➤ Rinse the pipette with Na_2CO_3 solution prepared in Part-I.

➤ Pipette out exactly 10 ml of the prepared Na_2CO_3 solution and transfer it to a clean conical flask.

➤ Add about add 2-3 drops of methyl orange indicator (the solution turns yellow)

➤ Titrate it against 0.1 M HC1 solution from the burette. When the solution becomes orange, stop adding HC1 solution and note the burette reading.

➤ Repeat the experiment till three Constant readings are obtained. Note the Mean Burette Reading (A).

Observation:

Burette	: 0.1 M HCl solution
Pipette	: 10 ml 'x' M Na_2CO_3 solution
Indicator	: Methyl orange
Colour change	: Yellow to orange
Equation	: $Na_2CO_3 + 2HCl \rightarrow 2NaCl + H_2O + CO_2$

Table

Burette Reading	Pilot	I	II	III	Mean Burette Reading
Final Reading					B. R. = _____ (A) ml
Initial Reading					
Difference					

Calculation:

The volume of 0.1 M HCl required for the neutralization of Na_2CO_3 = A ml

Molarity of Na_2CO_3 solution:

$$a\ M_1V_1 = b\ M_2V_2$$

Where,

M_1 = Molarity of HCl solution = 0.1M

V_1 = Volume of HCl solution = 'A' ml

M_2 = Molarity of Na_2CO_3 solution = 'X' M

V_2 = Volume of Na_2CO_3 solution = 10 ml

a = no. of moles of base = 1

b = no. of moles of acid = 2

☞ **gms / litre of Na_2CO_3 solution:**

gms/litre = Molarity x mol. Wt.

☞ **Normality of Na_2CO_3 solution:**

Normality = Molarity x molecular wt. / eq. wt. or

Normality = gms.litre^{-1} / eq. wt.

Result:

Solution	Eq. Wt	Mol. Wt	Molarity	gms/litre	Normalit
HC1	36.5	36.5			
Na_2CO_3	53	106			

Experiment No. 3

Aim: To determine the molarity, g/litre and normality of NaOH and HC1 using 0.05M

   Na_2CO_3 solution.

PART-1: To determine the molarity, g/litre and normality of HC1 using 0.05M Na_2CO_3 solution.

Requirements: 'x' M HC1, 0.05M Na_2CO_3 solution, methyl orange
Procedure:

➠ Wash all the apparatus with water.

➠ Rinse the burette with the given 0.05M Na_2CO_3 solution and fill the burette with the given 0.05 M Na_2CO_3 solution.

➠ Rinse the pipette with 'x' M HC1 solution.

➠ Pipette out exactly 10 ml of the given 'x' M HC1 solution and transfer to a clean conical flask.

➠ Add about 2-3 drops of methyl orange indicator (the solution turns orange)

➠ Titrate it against 0.05 M Na_2CO_3 solution from the burette.

When the solution becomes yellow, stop adding Na_2CO_3 solution and note the burette reading.

➡ Repeat the experiment till three constant readings are obtained. Note the Mean Burette Reading (A).

Observation:

Burette: 0.05 M Na_2CO_3 solution

Pipette: 10 ml 'x' M HC1 solution

Indicator: Methyl orange

Colour change: orange to yellow

Equation: $Na_2CO_3 + 2HC1 \rightarrow 2NaCl + H_2O + CO_2$

Table

Burette Reading	Pilot	I	II	III	Mean Burette Reading
Final Reading					B. R. = _____ (A) ml
Initial Reading					
Difference					

Calculation:

The volume of 0.05M Na_2CO_3 required for the neutralization of 'x' M HC1 solution = A ml

Molarity of HCl in the mixture:

$$a M_1 V_1 = b M_2 V_2$$

Where,

M_1 = Molarity of Na_2CO_3 solution = 0.05M

V_1 = Volume of Na_2CO_3 solution = 'A' ml

M_2 = Molarity of HCl solution = 'X' M

V_2 = Volume of HCl solution = 10 ml

a = no. of moles of acid = 2

b = no. of moles of base = 1

☞ **gms / litre of**
Na$_2$CO$_3$ in the
mixture:

gms/litre = Molarity x mol.
wt.

☞ **Normality of**
Na$_2$CO$_3$ in the
mixture:

Normality = Molarity x molecular wt. / eq. wt.

or

Normality = gms.litre'^{1} / eq. wt.

PART-2: To determine the molarity, g/litre and normality of 'y' M NaOH by using 'x' M HC1 solution.

Requirements: 'y' M NaOH solution, 'x' M HC1 solution, phenolphthalein

Procedure:

➤ Wash all the apparatus with water.

➤ Rinse the burette with the given xM HCI solution and fill the burette with the given xM HCI solution.

➤ Rinse the pipette with NaOH solution.

➤ Pipette out exactly 10 ml of the given NaOH solution and transfer to a clean conical flask.

➤ Add about add 2- 3 drops of phenolphthalein indicator (the solution turns pink).

➤ Titrate it against xM HCI solution from the burette. When the solution becomes colorless, stop adding HCI solution and note the burette reading.

➤ Repeat the experiment till three concordant readings are obtained. Note the Mean Burette Reading

(B).

Observation:

Burette: Y M HCI solution
Pipette: 10 ml y M NaOH solution
Indicator: Phenolphthalein
Colour change: pink to colorless
Equation:
NaOH + HCI->NaCl + H$_2$0

Table

Burette Reading	Pilot	I	II	III	Mean Burette Reading
Final Reading					B. R. = _____ (B) ml
Initial Reading					
Difference					

Calculation:

The volume of 0.1M HC1 required for the neutralization of 10 ml NaOH = B ml Molarity of NaOH solution:

$$a\ M_1V_1 = b\ M_2V_2$$

Where

M, = Molarity of HC1 solution = x M
V, = Volume of HC1 solution ='B' ml
M$_2$ = Molarity of NaOH solution = 'y' M
V$_2$ = Volume of NaOH solution = 10 ml
a = no. of moles of base = 1
b = no. of moles of acid = 1

☞ **gms / litre of NaOH solution :**
gm/litre = Molarity x mol. wt

☞ **Normality of NaOH solution:**
Normality = Molarity x molecular wt / eq. wt
or

Normality = gms.litre-1 / eq. wt

Result:

Solution	Eq. Wt.	Mol. Wt.	Molarity	gms/litre	Normalit
HC1	36.5	36.5			
NaOH	40	40			

Experiment No. 4

Aim : To determine the molarity, g/litre and normality of each component in a mixture of NaOH and Na_2CO_3 using 0.1 M HC1 solution.

Requirements: 0.1M HC1, mixture of NaOH and Na_2CO_3, Phenolphthalein, methyl orange

Procedure:

➤ Wash all the apparatus with water.

➤ Rinse the burette with the given 0.1M HC1 solution and fill the burette with the given 0.1M HC1 solution.

➤ Rinse the pipette with the given mixture of NaOH and Na_2CO_3 solution.

➤ Pipette out exactly 10 ml of the given mixture of NaOH and Na_2CO_3 solution and transfer to a clean conical flask. A

➤ Add about 2-3 drops of phenolphthalein indicator (the solution turns pink).

➤ Titrate it against 0.1M HCl solution from the burette. When the solution becomes colourless, stop adding HCl solution and note the burette reading (PR). [Do not throw the solution away] [At this stage from the mixture of NaOH and Na_2CO_3; the entire amount of NaOH and half the amount ofNa_2CO_3 is neutralized].

➤ To this colourless solution now add 2-3 drops of methyl orange (the solution turns yellow) and continue the titration against 0.1M HC1 from the burette. When the solution becomes orange, stop adding HC1 solution and note the burette reading (MR).

➤ Repeat the experiment till three concordant readings are obtained. Note the Mean Burette Reading (A).

Observation:

Burette: 0.1M HC1 solution

Pipette: 10 ml mixture of NaOH and Na_2CO_3 solution

Indicator: (1) Phenolphthalein (2) Methyl orange

Colour change: (1) Pink to colourless (2) Yellow to orange

Equation: $NaOH + HCl \longrightarrow NaCl + H_2O$

$Na_2CO_3 + HCl \longrightarrow NaHCO_3 + NaCl$

$NaHCO_3 + HCl \longrightarrow NaCl + H_2O + CO_2$

Table:

Burette Reading	I		II		III		Mean	
	PR	MR	PR	MR	PR	MR	PR	MR
Final								
Initial								
Difference								

i.e. The volume of 0.1 M HCl solution required to neutralize the entire amount NaOH + Na_2CO_3 in the mixture is (PR) ml and the volume of 0.l M HCl solution required to neutralize the remaining Vz Na_2CO_3 in the mixture is MR ml

Calculation:

Volume of 0.l M HCl solution required to neutralize the entire amount NaOH + $1/2$ Na_2CO_3 in the mixture = (PR) ml

Volume of 0.1M HCl solution required to neutralize the remaining $1/2$ Na_2CO_3 in the mixture = (MR) ml

Volume of 0.1M HCl solution required to neutralize NaOH in the mixture = (PR-MR) = 'A' ml

Volume of 0.1M HCl solution required to neutralize the entire amount of Na_2CO_3 in the mixture = (2MR) = 'B' ml

Molarity of NaOH in the mixture:

$$M_1V_1 x\ b = M_2V_2x\ a$$

Where,

M_1 = Molarity of HCl solution = 0.1M

V_1 = Volume of HCl solution = (PR-MR) = 'A' ml

M_2 = Molarity of NaOH solution = 'x' M

V_2 = Volume of NaOH solution (mixture) = 10 ml 'a'

= No. of moles of acid = 1

'b' = No. of moles of base = 1

☞ **gms/litre of NaOH in the mixture:**

gms/litre - Molarity x mol. wt.

☞ **Normality of NaOH in the mixture:**

Normality = Molarity x molecular wt / eq. wt

or

Normality = gms.litre-1 / eq. wt.

Molarity of Na_2CO_3 in the mixture:

$$M_1V_1 \times b = M_3V_3 \times a$$

Where,

M_1 = Molarity of HC1 solution = 0.1 M

V_1 = Volume of HC1 solution = (2MR) = 'B' ml

M_3 = Molarity of Na_2CO_3 solution = 'y' M

V_3 = Volume of Na_2CO_3 solution (mixture) = 10 ml

'a' = No. of moles of acid = 2

'b' = No. of moles of base = 1

☞ **gms/litre of Na_2CO_3 in the mixture:**

gms/litre = Molarity x mol. wt.

☞ **Normality of Na_2CO, in the mixture:**

Normality = Molarity x molecular wt. / eq. wt.

or

Normality = gms.litre-1 / eq. wt.

Result:

Soluti	Eq. Wt	Mol.	Molarit	gms/l	Normal
NaOH	40	40			
Na_2CO_3	53	106			

Experiment No. 5

Aim: To determine the molarity, g/litre and normality of each component in a mixture of $H_2C_2O_4.2H_2O$ and H_2SO_4 using 0.02 M $KMnO_4$ and 0.1 M NaOH solutions.

Chemical Requirements: 0.02 M $KMnO_4$ solution, mixture of $H_2C_2O_4.2H_2O$ and H_2SO_4 solution, 0.1M NaOH solution, dil. H_2SO_4 solution, indicator phenolphthalein.

Procedure:

Part I: Estimation of the mixture of $H_2C_2O_4.2H_2O$ and H_2SO_4 solution

➤ Wash all the apparatus with water.
➤ Rinse the burette with the given 0.1 M NaOH solution and fill the burette with the given NaOH solution.
➤ Rinse the pipette with the given mixture of $H_2C_2O_4$ H_2O and H_2SO_4 solution.
➤ Pipette out exactly 10 ml of the given mixture of $H_2C_2O_4.2H_2O$ and H_2SO_4 solution and transfer to a clean conical flask.
➤ Add 2-3 drops phenolphthalein indicator and titrate it against 0.1M NaOH solution from the burette. When the solution becomes light pink, stop adding NaOH solution and note the burette reading.
➤ Repeat the experiment till three constant readings are obtained. Determine the mean burette reading (A).

Part II: Determination of molarity, gm/litre and normality of $H_2C_2O_4.2H_2O$

➤ Wash all the apparatus with water.
➤ Rinse the burette with the given 0.02 M $KMnO_4$ solution and fill the burette with the given $KMnO_4$ solution.

➤ Rinse the pipette with the given mixture of $H_2C_2O_4.2H_2O$ and H_2SO_4 solution.

➤ Pipette out exactly 10 ml of the given mixture of $H_2C_2O_4.2H_2O$ and H_2SO_4 solution and transfer to a clean conical flask.

➤ Add about two test-tubes dil. H_2SO_4 solution and heat the flask to 60-70° C.

➤ Then immediately titrate it against 0.02 M $KMnO_4$ solution from the burette. When the solution becomes light pink, stop adding $KMnO_4$ solution and note the burette reading.

➤ Repeat the experiment till three concordant readings are obtained. Determine the mean burette reading (B).

Observation:

Part I: Estimation of the mixture of $H_2C_2O_4.2H_2O$ and H_2SQ_4 solution

Observation:

Burette: 0.1M NaOH solution

Pipette: 10 ml mixture of $H_2C_2O_4.2H_2O$ and H_2SO solution

Flask: 10 ml mixture of $H_2C_2O_4.2H_2O$ and H_2SO_4 solution + two test-tubes dil. H_2SO_4 solution and heat the flask to 60-70° C.

Indicator: Phenolphthalein

Colour Change: Colorless to pink

Equation:

$2NaOH + H_2SO_4 \rightarrow Na_2SO_4 + 2H_2O$

$2NaOH + H_2C_2O_4 \rightarrow Na_2C_2O_4 + 2H_2O$

$4NaOH + H_2SO_4 + H_2C_2O_4 \rightarrow Na_2SO_4 + Na_2C_2O_4 + 4H_2O$

Table:

Burette Reading	Pilot	I	II	III	Mean Burette Reading
Final Reading					B. R. = _____ (A) ml
Initial Reading					
Difference					

i.e. **'A'** ml 0.1 M NaOH solution is required to neutralize 10 ml 'mixture of $H_2C_2O_4.2H_2O$ and H_2SO_4 solution

Molarity of the mixture of $H_2C_2O_4.2H_2O$ and H_2SO_4 solution:

$$M_1V_1x\ b = M_3V_3x\ a$$

Where,

M_1= Molarity of mixture of $H_2C_2O_4.2H_2O$ and H_2SO_4 solution = Y M

V_1 = Volume of mixture of $H_2C_2O_4.2H_2O$ and H_2SO_4 solution = 10 ml

M_2 = Molarity of NaOH = 0.1 M

V_2 = Volume of NaOH = mean burette reading = 'A' ml

'a' = No. of moles of acid = 1

'b' = No. of moles of base = 2

Part II: Determination of molarity, g/litre and normality of $H_2C_2O_4.2H_2O$ in the mixture

Observation:

Burette: 0.02 M $KMnO_4$ solution

Pipette: 10 ml mixture of $H_2C_2O_4.2H_2O$ and H_2SO_4 solution

Flask: 10 ml mixture of $H_2C_2O_4.2H_2O$ and H_2SO_4 solution + two test-tubes dil H_2SO_4 solution and heat the flask to 60-70° C

Indicator: $KMnO_4$ self-indicator

Colour Change: Colorless to pink

Equation:

$2KMnO_4 + 3H_2SO_4 \rightarrow K_2SO_4 + 2MnSO_4 + 3H_2O + 5(O)$ [Oxidation]

$5H_2C_2O_4 + 5\ (O) \rightarrow 10\ CO_2 + 5H_2O$ [Reduction]

$2\ KMnO_4 + 5\ H_2C_2O_4 + 8H_2SO_4 \rightarrow K_2SO_4 + 2MnSO_4 + 10\ CO_2 + 8H_2O$

Ionic Equation:

$2MnO_4^- + 5C_2O_4^{2-} + 16H^+ \rightarrow 2Mn^{2+} + 10\ CO_2 + 8H_2O$

Table

Burette Reading	Pilot	I	II	III	Mean Burette Reading
Final Reading					B. R. = _____
Initial Reading					(A) ml
Difference					

i.e. 'B' ml 0.02M $KMnO_4$ solution is required to oxidize $H_2C_2O_4.2H_2O$ present in 10 ml mixture of $H_2C_2O_4.2H_2O$ and H_2SO_4 solution.

Calculations:
Molarity of $H_2C_2O_4.2H_2O$ solution:

$$M_1V_1 \times a = M_3V_3 \times b$$

Where,

M_1 = Molarity of $KMnO_4$ solution = 0.02M

V_1 = Volume of $KMnO_4$ solution = mean burette reading = 'B'ml

M_3 = Molarity of $H_2C_2O_4.2H_2O$ solution = 'y' M

V_3 = Volume of $H_2C_2O_4.2H_2O$ solution = 10 ml

'a' = No. of moles of electrons accepted by $KMnO_4$ = 5

'b' = No. of moles of electrons donated by $H_2C_2O_4.2H_2O$ = 2

☞ **gms / litre of $H_2C_2O_4.2H_2O$ solution:**

gms/litre = Molarity x mol. Wt.

☞ **Normality of $H_2C_2O_4.2H_2O$ solution:**

Normality = Molarity x molecular wt. / eq. wt.

Or

Normality = gms.litre^{-1} / eq. wt.

Molarity of H_2SO_4 in the mixture of $H_2C_2O_4.2H_2O$ and H_2SO_4 solution

Molarity of the mixture of $H_2C_2O_4.2H_2O$ and H_2SO_4 solution = 'x' M

Molarity of $H_2C_2O_4.2H_2O$ in the of the mixture of $H_2C_2O_4.2H_2O$ and H_2SO_4 solution = 'y' M

Molarity of H_2SO_4 in the of the mixture of $H_2C_2O_4.2H_2O$ and H_2SO_4 solution $=(x-y)$ M

☞ **gms / litre of H2S04 solution:**

gms/litre = Molarity x mol. wt

☞ **Normality of H_2SO_4 solution:**

Normality = Molarity x molecular wt. / eq. wt.

or

Normality = gms. Litre^{-1} / eq. wt.

Result:

Solution	Eq.wt.	Mol.	Molari	Norma	gms/
$H_2C_2O_4.2H_2O$	63	126			
H_2SO_4	49	98			

Experiment No. 6

Aim: To determine the molarity, g/litre and normality of each component in a mixture of $H_2C_2O_4.2H_2O$ and $K_2C_2O_4.H_2O$ using 0.02 M $KMnO_4$ and 0.1 M NaOH solutions

Chemical Requirements: 0.02 M $KMnO_4$, 0.1 M NaOH, mixture of $H_2C_2O_4.2H_2O$ and $K_2C_2O_4.H_2O$, dil. H_2SO_4 solution, phenolphthalein indicator

Procedure:

Part I: Determination of molarity of the mixture of $H_2C_2O_4.2H_2O$ and $K_2C_2O_4.H_2O$

➻ Wash all the apparatus with water.

➻ Rinse the burette with the given 0.02 M $KMnO_4$ solution and fill the burette with the given $KMnO_4$ solution.

➻ Rinse the pipette with the given mixture of $H_2C_2O_4.2H_2O$ and $K_2C_2O_4.H_2O$ solution.

➻ Pipette out exactly 10 ml of the given mixture of $H_2C_2O_4.2H_2O$ and $K_2C_2O_4.H_2O$ solution and transfer to a clean conical flask.

➻ Add about two test-tubes dil H_2SO_4 solution and heat the flask

60

to 60-70° C.

➡ Then immediately titrate it against $KMnO_4$ solution from the burette. When the solution becomes light pink, stop adding $KMnO_4$ solution and note the burette reading.

➡ Repeat the experiment till three concordant readings are obtained. Determine the mean burette reading (A).

Part II: Determination of Molarity, gm/litre and Normality of $H_2C_2O_4.2H_2O$ solution

➡ Wash all the apparatus with water.

➡ Rinse the burette with the given 0.1 M NaOH solution and fill the burette with the given NaOH solution.

➡ Rinse the pipette with the given mixture of $H_2C_2O_4.2H_2O$ and $K_2C_2O_4.H_2O$ solution.

➡ Pipette out exactly 10 ml of the given mixture of $H_2C_2O_4.2H_2O$ and $K_2C_2O_4.H_2O$ solution and transfer to a clean conical flask.

➡ Add 2-3 drops phenolphthalein indicator and titrate it against 0.l M NaOH solution from the burette. When the solution becomes light pink, stop adding NaOH solution and note the burette reading.

➡ Repeat the experiment till three concordant readings are obtained. Determine the mean burette reading (B).

Part I: Determination of molarity of the mixture of $H_2C_2O_4.2H_2O$ and $K_2C_2O_4.2H_2O$ solution:

Observation:

Burette: 0.02 M $KMnO_4$ solution

Pipette: 10 ml mixture of $H_2C_2O_4.2H_2O$ and $K_2C_2O_4.H_2O$ solution

Flask: 10 ml mixture of $H_2C_2O_4.2H_2O$ and $K_2C_2O_4.H_2O$ solution + two test-tubes dil. H_2SO_4 solution and heat the flask to 60- 70° C.

Indicator: $KMnO_4$ self-indicator

Colour Change: Colorless to pink

Equation:

$5K_2C_2O_4 + 5H_2SO_4$ -+ $5\ K_2SO_4 + 5H_2C_2O_4$

$2KMnO_4 + 3H_2SO_4$ -+ $K_2SO_4 + 2MnSO_4 + 3H_2O + 5(O)$ [Oxidation]

$5H_2C_2O_4 + 5\ (O)$ -+ $10\ CO_2 + 5H_2O$ [Reduction]

$2KMnO_4 + 5\ K_2C_2O_4 + 8H_2SO_4$ -+ $6^\wedge O,$ + $2MnSO_4 + 10\ CO_2 + 8H_2O$

Ionic Equation:

$2MnO_4^- + 5C_2O_4^{2-} + 16H^+ \longrightarrow 2Mn^{2+} + 10\ CO_2 + 8H_2O$

Table:

Burette Reading	Pilot	I	II	III	Mean Burette Reading
Final Reading					B. R. = _____
Initial Reading					
Difference					(A) ml

i.e. 'A' ml 0.02M $KMnO_4$ solution is required to oxidize 10 ml mixture of $H_2C_2O_4.2H_2O$ and $K_2C_2O_4.H_2O$ solution.

Calculations:

Molarity of mixture of $H_2C_2O_4.2H_2O$ and $K_2C_2O_4.H_2O$ solution:

$$M_1\ V_1\ x\ a = M_2\ V_2\ x\ b$$

Where

M_1 = Molarity of $KMnO_4$ solution = 0.02M

V_1 = Volume of $KMnO_4$ solution = mean burette reading = 'B' ml

M_2 = Molarity of mixture of $H_2C_2O_4.2H_2O$ and $K_2C_2O_4.H_2O$ solution = y M

V_2 = Volume of mixture $H_2C_2O_4.2H_2O$ and $K_2C_2O_4.H_2O$ solution = 10 ml

'a' = No. of moles of electrons accepted by $KMnO_4$ = 5

'b'= No. of moles of electrons donated by mixture ($H_2C_2O_4.2H_2O$ and $K_2C_2O_4.H_2O$) = 2

Part II: Determination of Molarity of $H_2C_2O_4.2H_2O$ solution

Observation:

Burette: 0.1 MNaOH solution

Pipette: 10 ml mixture of $H_2C_2O_4.2H_2O$ and $K_2C_2O_4.H_2O$ solution

Indicator: Phenolphthalein

Colour Change: Colourless to light pink

Equation:

$$H_2C_2O_4 + 2\ NaOH\ Na_2C_2O_4 + 2H_2O$$

Table:

Burette Reading	Pilot	I	II	III	Mean Burette Reading
Final Reading					B. R. = _____ (B) ml
Initial Reading					
Difference					

i.e. **'B'** ml 0.1 M NaOH solution is required to neutralize $H_2C_2O_4.2H_2O$ in 10 ml 'mixture of $H_2C_2O_4.2H_2O$ and $K_2C_2O_4.H_2O$ solution

Calculations:

Molarity of mixture of $H_2C_2O_4.2H_2O$

$$M_1\ V_1\ x\ a = M_2\ V_2\ x\ b$$

Where

M_1 = Molarity of NaOH solution = 0.02M

V_1 = Volume of NaOH solution = mean burette reading = 'B' ml

M_2 = Molarity of mixture of $H_2C_2O_4.2H_2O$ solution = y M

V_2 = Volume of mixture $H_2C_2O_4.2H_2O$ solution = 10 ml

'a'   = No. of moles of acid = 1

'b'   = No. of moles of base = 2

☞ **gms / litre of $H_2C_2O_4.2H_2O$ solution:**

$$gms/litre = Molarity\ x\ mol.\ Wt.$$

☞ **Normality of $H_2C_2O_4.2H_2O$ solution:**

63

Normality = Molarity x molecular wt. / eq. wt.

Or

Normality = gms.litre^{-1} / eq. wt.

Molarity of $K_2C_2O_4.H_2O$ in the mixture of $H_2C_2O_4.2H_2O$ and $K_2C_2O_4.H_2O$ solution:

Molarity of the mixture of $H_2C_2O_4.2H_2O$ & $K_2C_2O_4.H_2O$ solution = 'x" M

Molarity of $H_2C_2O_4.2H_2O$ in the of the mixture of $H_2C_2O_4.2H_2O$ & $K_2C_2O_4.H_2O$ solution = 'y' M

Therefore,

Molarity of $K_2C_2O_4.H_2O$ in the of the mixture of $H_2C_2O_4.2H_2O$ & $K_2C_2O_4.H_2O$ solution = (x-y) M

☞ **gms / litre of $K_2C_2O_4.H_2O$ solution:**

gms/litre = Molarity x mol. Wt.

☞ **Normality of $K_2C_2O_4.H_2O$ solution:**

Normality = Molarity x molecular wt. / eq. wt

Or

Normality = gms.litre^{-1} / eq. wt.

Result:

Solutio	Eq. Wt.	Mol.	Molarity	Normali	Gms/
$H_2C_2O_4.2H$	63	126			
$K_2C_2O_4.H_2O$	92	184			

Experiment No. 7

Aim: To determine the molarity, g/litre and normality of $KMnO_4$ and $FeSO_4.7H_2O$ solution using 0.05M $H_2C_2O_4.2H_2O$ solution.

Chemical Requirements: 0.05M $H_2C_2O_4.2H_2O$ solution, 'x' M $KMnO_4$ solution, y M $FeSO_4.7H_2O$ solution, dil. H_2SO_4 solution

Procedure:

Part I: Determination of molarity, g/litre and normality of $KMnO_4$ solution

➥ Wash all the apparatus with water.

➥ Rinse the burette with the given 'x' M $KMnO_4$ solution and fill the burette with the given $KMnO_4$ solution.

➥ Rinse the pipette with the given 0.05 M $H_2C_2O_4.2H_2O$ solution. Pipette out exactly 10 ml of the given 0.05 M $H_2C_2O_4.2H_2O$.

➥ Transfer to a clean conical flask. Heat the flask to 60-70° C. Then immediately titrate it against $KMnO_4$ solution from the burette. When the solution becomes light pink, stop adding $KMnO_4$ solution and note the burette reading.

➥ Repeat the experiment till three concordant readings are obtained. Determine the mean burette reading (A).

Part II: Determination of molarity, g/litre and normality of $FeSO_4.7H_2O$ solution

➥ Wash all the apparatus with water.

➥ Rinse the burette with the given 'x' M $KMnO_4$ solution and fill the burette with the given $KMnO_4$ solution.

➥ Rinse the pipette with the given *y* M $FeSO_4.7H_2O$ solution. Pipette out exactly 10 ml of the given $FeSO_4.7H_2O$ solution.

➥ Transfer to a clean conical flask. Add about two test-tubes dil H_2SO_4 solution and titrate it against Y M $KMnO_4$

solution from the burette. When the solution becomes light pink, stop adding $KMnO_4$ solution and note the burette reading.

➡ Repeat the experiment till three concordant readings are obtained. Determine the Mean Burette Reading (B).

Part I: Determination of molarity, g/litre and normality of $KMnO_4$ solution.

Observation:

Burette   : 'x'M $KMnO_4$ solution

Pipette   : 10 ml 0.05M $H_2C_2O_4.2H_2O$ solution

Flask   : 10 ml 0.05M $H_2C_2O_4.2H_2O$ solution + Heat the flask to 60-70° C.

Indicator : $KMnO_4$ self-indicator

Colour Change: Colorless to pink

Equation:

$2KMnO_4 + 3H_2SO_4 \rightarrow K_2SO_4 + 2MnSO_4 + 3H_2O + 5(O)$ [Oxidation]

$5 H_2C_2O_4 + 5 (O) \rightarrow 10 CO_2 + 5H_2O$ [Reduction]

$2KMnO_4 + 5 H_2C_2O_4 + 3H_2SO_4 \rightarrow K_2SO_4 + 2MnSO_4 + 10 CO_2 + 8H_2O$

Ionic Equation: $2MnO_4^- + 5C_2O_4^{2-} + 16H^+ \rightarrow 2Mn^{2+} + 10 CO_2 + 8H_2O$

Table:

Burette Reading	Pilot	I	II	III	Mean Burette Reading
Final Reading					B. R. = _____ (A) ml
Initial Reading					
Difference					

i.e. 'A' ml 'x'M $KMnO_4$ solution is required to oxidize 10 ml 0.05 M $H_2C_2O_4.2H_2O$ solution.

Calculations:
Molarity of 'x'M KMn04 solution:

$$M_1 V_1 \times a = M_2 V_2 \times b$$

Where

M_1 = Molarity of $KMnO_4$ solution = 'x' M

V_1 = Volume of $KMnO_4$ solution = mean burette reading = 'A' ml

M_2 = Molarity of mixture of $H_2C_2O_4.2H_2O$ and $K_2C_2O_4.H_2O$ solution = 0.05 M

V_2 = Volume of mixture $H_2C_2O_4.2H_2O$ solution = 10 ml

'a'   = No. of moles of electrons accepted by $KMnO_4$ = 5

'b'   = No. of moles of electrons donated by $H_2C_2O_4.2H_2O$ = 2

☞ **gms / litre of KMn04 solution:**

gms/litre = Molarity x mol. wt.

☞ **Normality of KMn04 solution:**

Normality = Molarity x molecular wt. / eq. wt.

Or

Normality = gms.litre^{-1} / eq. wt.

Part II: Determination of molarity, g/litre and normality of $FeSO_4.7H_2O$ solution:

Observation:

Burette:'x'M $KMnO_4$ solution

Pipette:10 m l'y' M $FeSO_4.7H_2O$ solution

Flask: 10 ml 'y' M $FeSO_4.7H_2O$ solution + 2 TT dil.H_2SO_4 solution

Indicator: $KMnO_4$ self-indicator

Colour Change: Colorless to pink

Equation:

$2KMnO_4 + 3H_2SO_4 \rightarrow K_2SO_4 + 2MnSO_4 + 3H_2O + 5(0)$ [Oxidation]

$10FeSO_4 + 5 H_2SO_4 + 5(0) \rightarrow 5Fe_2(SO_4)_3 + 5H_2O$ [Reduction]

$2KMnO_4 + 10FeSO_4 + 8H_2SO_4 \rightarrow K_2SO_4 + 2MnSO_4 + 5Fe_2(SO_4)_3 + 8H_2O$

Volumetric Analysis: Concepts and Experiments

Ionic Equation:

$$2MnO_4^- + 10Fe^{2+} + 16H^+ \rightarrow 2Mn^{2+} + 10Fe^{3+} + 8H_2O$$

Table:

Burette Reading	Pilot	I	II	III	Mean Burette Reading
Final Reading					B. R. = _____ (A) ml
Initial Reading					
Difference					

i.e. 'B' ml 'x'M $KMnO_4$ solution is required to oxidize 10 ml "y" M $FeSO_4.7H_2O$ solution.

Calculations:

Molarity of $FeSO_4.7H_2O$ solution:

$$M_1V_1x\ a = M_3V_3x\ b$$

Where,

M_1= Molarity of $KMnO_4$ solution = 'x' M

V_1= Volume of $KMnO_4$ solution = mean burette reading = 'B' ml

M_3= Molarity of $FeSO_4.7H_2O$ solution = 'y' M

V_3= Volume of $FeSO_4.7H_2O$ solution = 10 ml

'a' = No. of moles of electrons accepted by $KMnO_4$ = 5

'b' = No. of moles of electrons donated by $FeSO_4.7H_2O$ = 1

☞ **gms / litre of $FeSO_4.7H_2O$ solution:**

$$gms/litre = Molarity \times mol.\ wt$$

☞ **Normality of $FeSO_4.7H_2O$ solution:**

$$Normality = Molarity \times molecular\ wt / eq.\ wt$$

Or

$$Normality = gins.litre^{-1} / eq.\ wt$$

Result:

Solution	Eq. Wt.	Mol.	Molar	Normal	Gms/
$KMnO_4$	31.6	158			
$FeSO_4.7H_7O$	278	278			

Experiment No. 8

Aim: To determine the molarity, g/litre and normality of $FeSO_4(NH_4)_2SO_4.6H_2O$ and $K_2Cr_2O_7$ solutions using 0.02 M $KMnO_4$ solution.

Requirements: 0.02 M $KMnO_4$, 'x' M $FeSO_4(NH_4)_2SO_4.6H_2O$, 'y' M $K_2Cr_2O_7$, H_2SO_4, mixture of H_2SO_4 and Na_2HPO_4 (2:1), diphenylamine indicator.

Procedure:

Part I: Determination of Molarity, gm/litre and Normality of $FeSO_4(NH_4)_2SO_4.6H_2O$

➡ Wash all the apparatus with water.

➡ Rinse the burette with the given 0.02 M $KMnO_4$ solution and fill the burette with the given $KMnO_4$ solution.

➡ Rinse the pipette with the given 'x' M $FeSO_4(NH_4)_2SO_4.6H_2O$ solution. Pipette out exactly 10 ml of the given 'x' M $FeSO_4(NH_4)_2SO_4.6H_2O$ solution.

➡ Transfer to a clean conical flask. Add about two test-tubes dil H_2SO_4 solution and titrate it against 0.02 M $KMnO_4$ solution from the burette. When the solution becomes light pink, stop adding $KMnO_4$ solution and note the burette reading.

➡ Repeat the experiment till three concordant readings are obtained. Determine the Mean Burette Reading (A).

Part II: Determination of molarity, g/litre and normality of $K_2Cr_2O_7$ solution

➡ Wash all the apparatus with water.

➡ Rinse the burette with the given 'y' M $K_2Cr_2O_7$ solution and fill the burette with the given 'y' M $K_2Cr_2O_7$ solution.

➡ Rinse the pipette with the given 'x' M $FeSO_4(NH_4)_2SO_4.6H_2O$ solution. Pipette out exactly 10 ml of the given 'x'M $FeSO_4(NH_4)_2SO_4.6H_2O$ solution.

➡ Transfer to a clean conical flask. Add about two test-tubes mixture of H_2SO_4 and Na_2HPO_4 and 2-3 drops diphenylamine indicator.

➡ Then immediately titrate it against 'y' M $K_2Cr_2O_7$ solution from the burette. The solution first turns green and then violet; when the solution becomes violet, stop adding $K_2Cr_2O_7$ solution and note the burette reading.

➡ Repeat the experiment till three concordant readings are obtained. Determine the mean burette reading (B).

Part I: Determination of molarity, g/litre and normality of $FeSO_4(NH_4)_2SO_4.6H_2O$ solution:

Observation:

Burette   : 0.02 M $KMnO_4$ solution

Pipette   : 10 ml 'x' M $FeSO_4(NH_4)_2SO_4.6H_2O$ solution

Flask : 10 ml 'x' M $FeSO_4(NH_4)_2SO_4.6H_2O$ + two test tubes dil.H_2SO_4 Solution

Indicator: $KMnO_4$. self-indicator

Colour Change: Colourless to pink

Equation:

$2KMnO_4 + 3H_2SO_4 \rightarrow K_2SO_4 + 2MnSO_4 + 3H_2O + 5(O)$ [Oxidation]

$10FeSO_4(NH_4)_2SO_4 + 5H_2SO_4 + 5(O) \rightarrow 5Fe_2(SO_4)_3 + 10(NH_4)_2SO_4 + 5H_2O$ [Reduction]

$2KMnO_4 + 10FeSO_4(NH_4)_2SO_4 + 8H_2SO_4 \rightarrow K_2SO_4 + 2MnSO_4 + 5Fe_2(SO_4)_3 + 10(NH_4)_2SO_4 + 8H_2O$

Ionic Equation:

$2MnO_4' + 10Fe^{2+} + 16H^+ \rightarrow 2Mn^{2+} + 10Fe^{3+} + 8H_2O$

Table:

Burette Reading	Pilot	I	II	III	Mean Burette Reading
Final Reading					B. R. = _____ (A) ml
Initial Reading					
Difference					

i.e. 'A' ml 0.02 M $KMnO_4$ solution is required to oxidize 10 ml 'x' M $FeSO_4(NH_4)_2SO_4.6H_2O$ solution.

Calculations:

Molarity of $FeSO_4(NH_4)_2SO_4.6H_2O$ solution:

$$M_1V_1 \times a = M_2V_2 \times b$$

Where

M_1 = Molarity of $KMnO_4$ solution = 0.02 M

V_1 = Volume of $KMnO_4$ solution = mean burette reading = 'A' ml

M_2 = Molarity of $FeSO_4(NH_4)_2SO_4.6H_2O$ solution = Y M

V_2 = Volume of $FeSO_4(NH4)_2SO_4.6H_2O$ solution = 10 ml

'a' = No. of moles of electrons accepted by $KMnO_4$ = 5

'b' = No. of moles of electrons donated by $FeSO_4(NH_4)_2SO_4.6H_2O$ = 1

☞ **gms / litre of $FeSO_4(NH_4)_2SO_4.6H_2O$ solution:**

gms/litre = Molarity x mol. wt.

☞ **Normality of $FeSO_4(NH_4)_2SO_4.6H_2O$ solution:**

Normality = Molarity x molecular wt. / eq. wt.

or

Normality = $gms.litre^{-1}$ / eq. wt.

Part II: Determination of molarity, g/litre and normality of $K_2Cr_2O_7$ solution

Observation:

Burette: 'y' M $K_2Cr_2O_7$ solution

Pipette: 10 ml Y M $FeSO_4(NH_4)_2SO_4.6H_2O$ solution

Flask : 10 ml Y M $FeSO_4(NH_4)_2SO_4.6H_2O$ + 2 test tubes mixture of H_2SO_4 and Na_2HPO_4.

Indicator: Diphenylamine

Colour Change : Colourless to violet

Equation　　　　　:

$K_2Cr_2O_7 + 4H_2SO_4 \rightarrow K_2SO_4 + Cr_2(SO_4)_3 + 4H_2O + 3(O)$ [Oxidation]

$6FeSO_4(NH_4)_2SO_4 + 3H_2SO_4 + 3(O) \rightarrow 3Fe_2(SO_4)_3 + 6(NH_4)_2SO_4 + 3H_2O$ [Reduction]

$K_2Cr_2O_7 + 6FeSO_4(NH_4)_2SO_4 + 7H_2SO_4 \rightarrow K_2SO_4 + Cr_2(SO_4)_3 + 3Fe_2(SO_4)_3 + 6(NH_4)_2SO_4 + 7H_2O$ [Redox]

Ionic Equation:

$Cr_2O_7^{2-} + 6Fe^{2+} + 17H^+ \rightarrow 20Cr^{+3} + 6Fe^{3+} + 7H_2O$

Table:

Burette Reading	Pilot	I	II	III	Mean Burette Reading
Final Reading					B. R. = _____
Initial Reading					(A) ml
Difference					

i.e. 'B' ml 'y'M $K_2Cr_2O_7$ solution is required to oxidize 10 ml 'x'M $FeSO_4(NH_4)_2SO_4.6H_2O$ solution.

Calculations:

Molarity of $K_2Cr_2O_7$ solution:

$$M_2V_2xa = M_3V_3x\ b$$

Where,

M_2 = Molarity of $FeSO_4(NH_4)_2SO_4.6H_2O$ solution = 'y' M

V_2 = Volume of $FeSO_4(NH_4)_2SO_4.6H_2O$ solution = 10 ml

M_3 = Molarity of $K_2Cr_2O_7$ solution = 'y'M

V_3 = Volume of $K_2Cr_2O_7$ solution = mean burette reading = 'B' ml

'a' = No. of moles of electrons donated by $FeSO_4(NH_4)_2SO_4.6H_2O$ = 1

'b' = No. of moles of electrons accepted by $K_2Cr_2O_7$ = 6

☞ **gms / litre of $FeSO_4(NH_4)_2SO_4.6H_2O$ solution:**

gms/litre = Molarity x mol. wt

☞ **Normality of $FeSO_4(NH_4)_2SO_4.6H_2O$ solution:**

Normality = Molarity x molecular wt. / eq. wt.

or

Normality = gms.litre"^{1} / eq. wt.

Result:

Solution	Eq.	Mol.	Molarity	Normality	Gms/
$FeSO_4(NH_4)_2SO_4.6H_2O$	392	392			
$K_2Cr_2O_7$	49	294			

Exercise for Entry Level Students
Level – 2

Experiment No. 1

Aim : To determine the amount of Cu^{+2} in given solution of $CuCl_2.2H_2O$ using 0.01 M EDTA

Requirements :
0.01 M EDTA, 6 N NH_4OH , unknown $CuCl_2.2H_2O$, distilled water.

Principle :
EDTA (disodium salt) is used as titrant. EDTA $(Na_2H_2C_{10}H_8N_2.2H_2O)$ is given as Na_2H_2Y. Hence in aqueous solution it given H_2Y^{-2} complex ion, when it reacts with metal, it gives reaction given below.

$$M^{+2} + H_2Y^{-2} \rightarrow MY^{-2} + 2H^+$$

Procedure :
- Wash the burette with Distilled water.
- Rinse with EDTA (0.01M) and remove air trapped in burette if any.
- Dilute given unknown solution of $CuCl_2.2H_2O$ upto 250 ml with distilled water, shake well.
- Pipette out 25 ml of $CuCl_2.2H_2O$ in a conical flask
- Add 2 ml 6 N NH_4OH
- Add 3-4 drops of FSB-F indicator
- Swirl well the conical flask.
- Titrate against 0.01M EDTA soln.
- At the end point color change will appear from Blue/Black/Purple (according to chemical purity) to green
- Note this reading as pilot (Range)
- Repeat the procedure three more times.

Observation:

Burette   : 0.01 M EDTA

Conical Flask: 25 ml diluted solution of $CuCl_2 . 2H_2O$ + 2 ml 6 N NH_4OH + 3 to 4 drops of indicator.

Indicator   :   Fast sulphon Black - F

Color change: Blue / Black / Purple to Green

Observation Table:

Initial	Pilot Reading	I	II	III	Constant
Final					
Burette					A=_____
Difference					

Chemical Equation :

$$MI_n + H_2Y^{-2} \rightarrow MY^{-2} + In^{-2} + 2H^+$$

Calculation :

1000 ml 1M EDTA $\cong$ 63.6 gm Cu^{+2}

1ml 1M EDTA $\cong$ 0.0636 gm Cu^{+2}

1 ml 0.01M EDTA $\cong$ 0.000636 gm Cu^{+2}

In 25 ml sample solution $\cong$ 0.000636 $\times$ A gm Cu^{+2}

$\qquad\qquad \cong$ gm Cu^{+2} (B)

In 250 ml sample solution $\cong$ B $\times$ 10 =_____ gm Cu^{+2} in given sol^n (C)

Result :

(1)   Vol. of 0.01 M EDTA for 25 ml solution = _____ (A)

(2)   Amount of Cu in given solution = _____ (C)

◆ Experiment No. - 2 ◆

Aim :

Estimation of the amount of Ni in the given solution of $NiSO_4.7H_2O$ using 0.01 M EDTA solution.

Requirement :

0.01 M EDTA, 7 pH buffer, unknown sol^n of $NiSO_4.7H_2O$, Murexide indicator.

Principle :

EDTA (disodium salt) is used as titrant. EDTA $(Na_2H_2C_{10}H_8N_2.2H_2O)$ is given as Na_2H_2Y. Hence in aqueous solution it given H_2Y^{-2} complex ion, when it reacts with metal, it gives reaction as given below.

$$M^{2+} + H_2Y^{2-} \rightarrow MY^{2-} + 2H^+$$

Procedure :

➹ Wash the burette with distilled water (D.W.)

➹ Rinse the burette with 0.01 M EDTA and remove the air trapped in burette if any.

➹ Dilute given solution of $NiSO_4.7H_2O$ with D.W. up to 250 ml in a measuring flask.

➹ Pipette out 25 ml of diluted $NiSO_4.7H_2O$ sol^n and add it in a conical flask.

➹ Add 50 ml distilled water (D.W.)

➹ Add about 100 mg indicator mixture. (50 mg Murexide + 50 mg KNO_3)

➹ Swirl well to dissolve indicator

➹ Add 10 ml 7 pH buffer solution.

➹ Titrate using 0.01 M EDTA

➹ At the end point color changes from yellow to purple.

➹ Note this reading as pilot reading

➹ Repeat this procedure three more times.

Observation :

Burette : 0.01 M EDTA

Conical flask: 25 ml diluted $NiSO_4.7H_2O$ + 50 ml D.W. (approx)
+ 10 ml 7 pH buffer sol^n + 100 mg indicator (solid mixture)

Indicator : (50 mg Murexide + 50 mg KNO_3) solid mixture

Color change : Yellow to Purple

Observation Table:

Initial	Pilot Reading	I	II	III	Constant
Final					
Burette					A=_____
Difference					

Chemical equation :

$$Ni^{+2}H2Y^{-2} \rightarrow NiY^{-2} + 2H^+$$

Calculation :

1000 ml 1M EDTA $\cong$ 58.69 gm Ni^{+2}

1ml 1M EDTA $\cong$ 0.05869 gm Ni^{+2}

1 ml 0.01M EDTA $\cong$ 0.0005869 gm Ni^{+2}

In 25 ml sample solution $\cong$ 0.0005869 $\times$ A gm Ni^{+2}

$\cong$ _____ gm Ni^{+2} (B)

In 250 ml sample solution $\cong$ B $\times$ 10

= _____ gm Ni^{+2} in given solution (250 ml)

Result :

1. Volume of 0.01 M EDTA required for 25 ml = _____ ml (A)

2. Amount of Ni in given solution = _____ gm (C)

◆ Experiment No. - 3 ◆

Aim :

Estimation of the amount of Zn present in the given solution of $ZnCl_2$ using 0.01 M EDTA solution.

Requirement :

0.01 M solution of EDTA, 10 pH buffer solution, unknown solution of $NiSO_4.7H_2O$, Eriochrome Black-T (EBT) indicator.

Principle :

EDTA (disodium salt) is used as titrant. EDTA $(Na_2H_2C_{10}H_8N_2.2H_2O)$ is given as Na_2H_2Y. Hence in aqueous solution it given H_2Y^{-2} complex ion, when it reacts with metal, it gives reaction given below.

$$M^{+2} + H_2Y^{-2} \rightarrow MY^{-2} + 2H^+$$

Procedure :

➤ Wash the burette with D.W.

➤ Rinse the burette with 0.01 M EDTA and remove the air trapped in burette if any.

➤ Dilute given solution of $NiSO_4.7H_2O$ with D.W. upto 250 ml in a measuring flask.

➤ Pipette out 25 ml of diluted $NiSO_4.7H_2O$ solution and add it in a conical flask

➤ Add 50 ml distilled water (D.W.)

➤ Add 2 ml 10 pH buffer solution

➤ Add 2-3 drops of E.B.T. indicator.

➤ Swirl well and titrate against 0.01 M EDTA $so1^n$

➤ At the end point color changes from wine red to blue.

➤ Note this reading as pilot reading (Range)

➤ Repeat this procedure three more times.

Observation Table:

Initial	Pilot Reading	I	II	III	Constant
Final					
Burette					A=_____
Difference					

Observation:

Burette : 0.01 M EDTA

Conical Flask: 25 ml diluted solution of Zn + 2 ml 10 pH buffer +2-3 drops of EBT indicator

Indicator : Eriochrome Black - T

Color change: Wine red to Blue

Chemical Equation :

$$Zn^{+2} + H_2In^{-1} \rightarrow ZnIn^- + 2H^+$$

$(Indicator)$ $\quad$ $(Win\,Red)$

$$ZnIn^- \rightarrow ZnY^{-2} + HIn^- + H^+$$

$(Wine\,Red)$ $\quad$ $(Blue)$

Calculation :

1000 ml 1M EDTA $\cong$ 65.38 gm Zn^{+2}

1ml 1M EDTA $\cong$ 0.06538 gm Zn^{+2}

1 ml 0.01M EDTA $\cong$ 0.0006538 gm Zn^{+2}

In 25 ml sample solution $\cong$ 0.0006538 × A gm Zn^{+2}

$\cong$ _____ gm Zn^{+2} (B)

In 250 ml sample solution $\cong B \times 10$

= _____ gm Zn^{+2} in given solution sol^n (C)

Result :

(1) Volume of 0.01 M EDTA required for 25 ml solution = _____ ml (A)

(2) Amount of Zn in given solution = _____ gm (C)

◀ Experiment No. - 4 ▶

Aim :

Determination of the acetic acid in commercial vinegar using 0.1 M $NaOH$

Requirement: Vinegar, 0.1 N $NaOH$, phenolphthalein as an indicator

Procedure:

➡ Dilute given solution of vinegar to 250 ml in a measuring flask with D.W.

➡ Transfer 25 ml to conical flask with pipette.

➡ Add 2-3 drops of phenolphthalein as an indicator.

➡ Titrate against 0.1 N $NaOH$ using burette.

➡ At the end point color of the solution will change from colorless to pink.

➡ Note the burette reading

➡ Repeat this procedure three more time.

Observation:

Burette   : 0.01 N EDTA

Conical Flask: 25 ml diluted solution of vinegar + 2-3 drop indicator

Indicator: Phenolphthalein

Color change: Colorless to light pink

Observation Table:

Initial	Pilot Reading	I	II	III	Constant
Final					
Burette					A=_____
Difference					

✿ **Chemical equation :**

$CH_3COOH + NaOH \rightarrow CH_3COONa + H_2O$

(vinegar)

✿ **Calculation :**

1000 ml 1N $NaOH$ = 60 gm CH_3COOH

1ml 1N $NaOH$ = 0.06 gm CH_3COOH

1 ml 0.1N $NaOH$ = 0.006 gm CH_3COOH

In 25 ml sample solution = $0.006 \times$ CH_3COOH gm (A)

$$= \underline{\hspace{2cm}} CH_3COOH \text{ gm (B)}$$

In 250 ml vinegar solution = B $\times 10$

$$= \underline{\hspace{2cm}} CH_3COOH \text{ gm (C)}$$

In 1000 ml vinegar sol^n = C $\times$ 4

$$= \underline{\hspace{2cm}} \text{gm / lit. (D)}$$

Result :

(1) Amount of CH_3COOH in given sol^n = _____ gm

(2) Amount of CH_3COOH in commercial vinegar = ____ gm / lit.

◆ Experiment No. - 5 ◆

Aim :

Estimation of total hardness of water by EDTA

Requirements :

0.1 M EDTA solution, 10 pH buffer solution, Eriochrome Black-T indicator, double distilled water.

Principle :

➡ Hard water contains dissolved salts of calcium and magnesium in the form of bicarbonates, sulphates and chlorides. Hardness of water is of two types;

  (i) Temporary Hardness

  (ii) ii) Permanent Hardness.

➡ Hardness is sum of calcium and magnesium which is expressed as $CaCO_3$ in parts per million (ppm) for both Ca^{+2} & Mg^{+2})

Procedure:

➡ Take 25 ml of sample in a 250 ml measuring flask and dilute it upto the mark with double distilled water.

➡ Pipette out 25 ml water sample in a conical flask.

➡ Add 3 ml 10 pH buffer solution.

➡ Add 2-3 drops of EBT indicator.

➡ Swirl well the conical flask.

➡ Titrate against 0.01 M EDTA.

➡ At the end point color change will appear from wine red to blue.

➡ Note this reading as pilot reading (Range).

➡ Repeat the procedure three more times.

Observation:

Burette: 0.01 M EDTA

Conical flask: 25 ml water sample + 3 ml 10 pH buffer + 2-3 drop EBT indicator.

Indicator: Eriochrome Black-T

Color change: Wine Red to Blue

Observation Table :

Initial	Pilot Reading	I	II	III	Constant
Final					
Burette					A=_____mL
Difference					

Chemical equation :

(i) $\quad CaR + Na_2C_{10}H_{14}O_8N_2 \rightarrow CaC_{10}H_{14}O_8N_2 + Na_2R$

(ii) $\quad MgR + Na_2C_{10}H_{14}O_8N_2 \rightarrow MgC_{10}H_{14}O_8N_2 + Na_2R$

$$[\text{Where } R = HCO_3^-, Cl^- \text{ or } SO_4^{-2}]$$

Calculation :

1000 ml 1M EDTA $\cong$ 100 gm $CaCO_3$

1ml 1M EDTA $\cong$ 100 gm $CaCO_3$

1 ml 0.01M EDTA $\cong$ 1 gm $CaCO_3$

In 25 ml sample solution $\cong$ $1 \times A$ gm $CaCO_3$

$$= B \text{ gm } CaCO_3$$

25 ml sample = B mg

1000 ml sample $\quad = ?$

$$\therefore \quad \frac{1000 \times B}{25} = \underline{\quad\quad} mg/mL(ppm)$$

Result :

(1) Total hardness $(as\ CaCO_3)$ of given sample water = ____ ppm

◆ Experiment No. - 6 ◆

Aim:

Determination of alkali present in antacid using 0.1 N HCl

Requirement:

Antacid tablet, 1 N HCl, 0.1 N NaOH, phenolphthalein indicator.

Part-1

Procedure :

➤ Dissolve one tab of antacid in 25 ml 1 N HCl solution and then dilute this solution to 250 ml in a measuring flask.

➤ Shake well before use.

➤ Take 25 ml of antacid solution using pipette and transfer it in conical flask.

➤ Add 2-3 drops of phenolphthalein indicator.

➤ Titrate against 0.1 N NaOH (from burette).

➤ At the end point color change will be observed from colorless to pink.

➤ Record this reading and repeat this procedure for three more times.

Observation:

Burette: 0.1 N NaOH

Conical Flask: 25 ml diluted antacid solution + 2-3 drops of indicator

Color change: Colorless to pink

Observation Table :

Initial	Pilot Reading	I	II	III	Constant
Final					
Burette					A=_____
Difference					

Part-2

Procedure:

➤ Take exact 25 ml of 1 N HCl in a measuring flask using pipette and make upto 250 with D.W.

➤ Pipette out 25 ml from diluted solution and transfer it to conical flask

➤ Add 1-2 drops of phenolphthalein indicator.

➤ Titrate against 0.1 N NaOH (from burette).

➤ At the end point color will change from colorless to pink.

➤ Record burette reading.

➤ Repeat this process 3 more times.

Observation:

Burette: 0.1 N NaOH

Conical Flask: 25 ml diluted HCl + 1-2 drops of indicator

Indicator: Phenolphthalein

Color change: Colorless to pink

Observation Table:

Initial	Pilot Reading	I	II	III	Constant
Final					
Burette					A=_____
Difference					

Chemical Reaction :

$$Mg(OH)_2 + 2HCl \rightarrow NgCl_2 + 2H_2O$$

$$NaOH + HCl \rightarrow Nacl + H_2O$$

Calculation :

☞ **Note :**

✳ Part-I of the practical gives reading for remaining HCl in the solution after the neutralization reaction with alkali from antacid.

✳ Whereas part-II gives reading for total HCl only, as there is no alkali present in conical flask.

Here in this practical alkali is represented as $Mg(OH)_2$ from amount of $Mg(OH)_2$ in sample,

1000 ml 1 m NaOH = 1000 ml 1 M HCl $= Y_2 \times 1000 ml$ $1M\ Mg(OH)_2$

1000 ml 1 m NaOH = 29.15 gm $Mg(OH)_2$ (Became M.W. of $Mg(OH)_2$ = 58.30)

1000 ml 1 M NaOH = 0.02915 gm $Mg(OH)_2$

1ml 0.1 M NaOH = 0.002915 gm $Mg(OH)_2$

A ml 0.1 M NaOH = 0.002915 × A gm

∴ Amount of Alkali present in 25 ml

dil. Antacid sol^n = 0.002915 × A gm

= _____ gm (B)

∴ Amount of total alkali present in given

dil. Antacid sol^n = 0.002915 × A × 10 gm

= _____ gm (C)

◆ Experiment No. - 7 ◆

Aim:

To determine the amount of Fe^{+3} in given solution of $FeCl_3.6H_2O$ (acidic) by reduction method.

Principle:

In presence of HCl & $SnCl_2$, $FeCl_3$ can be reduced to $FeCl_2$ and there after it can be titrated against $K_2Cr_2O_7$ using diphenylamine as an indicator.

Requirements:

Solution of $FeCl_3.6H_2O$, 0.1 N $K_2Cr_2O_7$, $HgCl_2$ solution, dilute H_2SO_4, Na_2HPO_4 solution, diphenylamine indicator, distilled water.

Procedure:

➻ Take 25 ml of $FeCl_3.6H_2O$ solution and dilute it up to 250 ml with D.W. using standard measuring flask.

➻ Transfer 25 ml of this diluted solution to a conical flask.

➻ Add 5 ml con. HCl and heat it up to 70^0C.

➻ To reduce Fe^{+3} in Fe^{+2} add $SnCl_2$ drop wise from burette while the solution is hot till it become colorless from yellow. Note this reading as 'X' ml.

➻ Allow to cool this reaction mixture in the same conical flask at room temp.

➻ Then add a few drops of $HgCl_2$. (Solution will become milky on addition of $HgCl_2$).

➻ Now add 10 ml (approx) distilled water.

➻ Add dil. H_2SO_4 10 ml (Approx one test tube).

➻ Add Na_2HPO_4 5 ml (Approx half test tube).

➻ Add 2-3 drops of diphenylamine as an indicator.

➤ Titrate against 0.1 N $K_2Cr_2O_7$.

➤ Note this burette reading and repeat the same procedure for three more times.

☞ **Note :**

✳ If solution does not become milky after the addition of $HgCl_2$ (Instep-6), do not continue the titration, discard solution and restart from step-1.

✳ If the solution becomes black in step-4 after the step-5, then also discard the solution and restart from step-1.

Chemical Reaction:
Observation:

$$2FeCl_3 + SnCl_2 \rightarrow 2FeCl_2 + SnCl_4$$
$$K_2Cr_2O_7 + 4H_2SO_4 \rightarrow K_2SO_4 + Cr_2(SO_4)_3 + 4H_2O + 3(O)$$
$$6FeCl_2 + 6HCl + 3(O) \rightarrow 6FeCl_3 + 3H_2O$$
$$SnCl_2 + 2HgCl_2 \rightarrow Hg_2Cl_2 + SnCl_4$$

Observation :

Burette: 0.1 N $K_2Cr_2O_7$

Conical Flask: 25 ml diluted $FeCl_3.6H_2O$ + 5 ml con. HCl + $SnCl_2$ (x ml) + $HgCl_2$ (2-4 drops) + H_2O + dil. H_2SO_4 + Na_2HPO_4

Indicator: Diphenylamine

Color change: Milky white → green → purple.

❀ **Observation Table :**

Initial	Pilot Reading	I	II	III	Constant
Final					
Burette					A=_____
Difference					

Calculation:

1000 ml 1 N $K_2Cr_2O_7$ = 270.3 gm $FeCl_3.6H_2O$ = 55.8 gm Fe

$$1 \text{ ml } 0.1 \text{ N } K_2Cr_2O_7 = \frac{270.3}{1000} \text{ gm } FeCl_3.6H_2O = \frac{A \times 55.8}{10000}$$

$\therefore$ 25 ml dill. $FeCl_3.6H_2O$ = A$\times$ 0.02703 gm $FeCl_3.6H_2O$ gm

Amount of $FeCl_3.6H_2O$ in given Sol^n = A $\times$ 0.02703 $\times$ 10 $FeCl_3.6H_2O$

$$= B \text{ gm } FeCl_3.6H_2O$$

$\therefore$ Amount of Fe^{+3} present in dil. 25 ml of $FeCl_3.6H_2O$ = A$\times$ 0.00558 gm

$\therefore$ In given solution, amount of Fe = 0.0058 $\times$ A$\times$ 10

$$= C \text{ gm}$$

✿ **Result :**

(1) Vol. of 0.1 N $K_2Cr_2O_7$ required to titrate 25 ml dil. $FeCl_3.6H_2O$ = (A) ml (A)

(2) Amount of $FeCl_3.6H_2O$ in given Sol^n = (B) gm

(3) Amount of Fe^{+3} in given solution = (C) gm